AUSTRALIA
THE CONTINENT OF EXTREMES
OUR GEOGRAPHICAL RECORDS

AUSTRALIA
THE CONTINENT OF EXTREMES
OUR GEOGRAPHICAL RECORDS

IAN READ

AUSTRALIA A CONTINENT
OF EXTREMES
OUR GEOGRAPHICAL RECORDS
by Ian Read

ISBN 1863152377

Designer: Dan Hormillosa
Cover design: Darren Holt
Edited and assisted by: Angus Paterson
Publisher: Charles Burfitt

Little Hills Press
Unit 3/18 Bearing Road
Seven Hills, NSW 2147
www.littehills.com
info@littlehills.com

Other books by the author
Australian Road Trips, Little Hills Press
Camping + Caravanning Across Australia, Little Hills Press
Doing the West Coast, Little Hills Press
The Bush: A Guide to the Vegetated Landscapes of Australia, UNSW Press, Sydney

Acknowledgments see page 150

Disclaimer
While all attempts have been made to insure the accuracy of the information herein things change over time and some of the facts listed may well be exceeded during the currency of this book. The listings herein are believed to be the extremes, as stated, based on current available information. Where doubts exist they have been noted in the text. Nevertheless, some inaccuracies may occur and sometimes value judgements have had to be made.
Any further information is gladly received and may be included in further editions of this book.

CONTENTS

List of Tables 7
List of Figures 10
List of Symbols and Abbreviations 10

Chapter One: Discovery and Settlement 11
   Aboriginal Occupation
   Discovery and Exploration
   Early European Settlement
   Old Australian Buildings
   Tall Australian Buildings

Chapter Two: Australia at Large 27
   The Australian Landmass
   Gondwana Legacy
   The Human Dimension - A Widespread
      Settled Population
   Australia's Plains
   The Human Dimension - Life on the
      Plains
   Australian Deserts
   The Human Dimension -
      Crossing the Desert

THE LAND

Chapter Three: The Land: Hills, Mountains and Tablelands 45
   Creation of Landscapes
   Mountain Building
   Tablelands
   The Human Dimension -
      Occupying the High Country
   Volcanoes
   Earthquakes

Chapter Four: The Land: Gorges and Valleys, Caves and Craters 61
   Gorges and Valleys
   Hillslopes and Cliffs
   The Human Dimension - Crossing the
      Range and Bridging the Gap
   Caves
   Big Holes
   Impact Craters

THE WATER

CHAPTER FIVE: Streams, Waterfalls and Lakes 75
- Streams
- The Human Dimension - Crossing the Bridge
- Waterfalls
- Lakes
- The Human Dimension - Some Hydrological Works
- Hot Springs

THE COAST

CHAPTER SIX 89
- The Coastline
- Gulfs and Bays
- Beach Coasts
- Rock Coasts
- Tidal Plain Coasts
- Coral Coasts
- Islands
- Straits
- Tides

THE BUSH

CHAPTER SEVEN 101
- Classification of Vegetation
- Australian Vegetation Types
- Changes to Vegetation Formations
- Vegetation - the Human Dimension

THE SKY

CHAPTER EIGHT 115
- Weather and Climate
- Day-to-Day Weather Changes
- Season-to-Season Weather Changes
- Extreme Temperatures
- Extreme Rain Events
- The Human Dimension - Impacts of Extreme Weather Conditions

Figures 138
Appendix: Unusual Placenames 148
Reference and Acknowledgements 150
Index 151

LIST OF TABLES

EXPLORATION AND SETTLEMENT
1.1 First Contacts between Aborigines and Europeans
1.2 Exploration and Discovery of Australian Landmass: Before 1788
1.3 First Temporary European Settlement in each State/Territory
1.4 Oldest European Settlements in Australia
1.5 Oldest European Settlement in each State/Territory
1.6 Some of Australia's Oldest Buildings
1.7 Oldest Building in each State/Territory
1.8 Oldest House in each State/Territory
1.9 Oldest Farmhouse or Homestead in each State/Territory
1.10 Oldest Public or Utilitarian Building in each State/Territory
1.11 Oldest Church in each State/Territory
1.12 Oldest Hotel Building in each State/Territory
1.13 Some Old Continuous Hotel Licenses in Australia
1.14 Australia's Oldest Bridges
1.15 Oldest Road Bridge in each State/Territory
1.16 Oldest Railway Bridge in each State/Territory
1.17 Oldest Mines in each State/Territory
1.18 Australia's Tallest Buildings
1.19 Tallest Building in each State/Territory
1.20 Some Tall Towers and Structures

AUSTRALIA AT LARGE
2.1 Land Areas of each State/Territory
2.2 Some Facts about Australia
2.3 Australia's Extreme Geographical Coordinates
2.4 Extreme Geographical Coordinates of each State/Territory
2.5 Australia's Extreme Sitings of Settlement
2.6 Extreme Sitings of Settlements for States/Territories
2.7 Coordinates of the Capital Cities
2.8 Australia's Most Inland Settlements
2.9 Most Inland Settlement in each State/Territory
2.10 Australia's Remotest Settlements or Stations
2.11 Most Isolated Settlement in each State/Territory
2.12 Most Isolated Farm or Station in each State/Territory
2.13 Australia's Longest Highways
2.14 Some Long Walking Trails
2.15 Some Australian Plains: by size
2.16 Straightest Road in each State/Territory
2.17 Straightest Section of Railway in each State/Territory
2.18 Some Long Australian Pipelines
2.19 Australian Deserts: by size
2.20 Australian Desert Tracks
2.21 Some Famous Outback Roads

THE LAND: HILLS, MOUNTAINS AND TABLELANDS
3.1 Highest Mountains in each State/Territory
3.2 The '7000 footers': Peaks and Tops over 7000 feet in New South Wales
3.3 The '6000 footers': Peaks and Tops over 6000 feet in Victoria
3.4 Highest Mountains: Other States/Territories
3.5 Highest Mountains: Australian Islands
3.6 Heights of some Well-known Peaks
3.7 Highest Located Structure/Building in each State/Territory
3.8 Australia's Highest Settlements
3.9 Australia's Highest Towns and Townships
3.10 Highest Settlement in each State/Territory
3.11 Highest Town or Township in each State/Territory
3.12 Highest Suburb of Capital Cities in each State/Territory

3.13 Highest Farm or Station In Each State/Territory
3.14 Highest Roads in each State/Territory
3.15 Highest Highway in each State/Territory
3.16 Highest Railway in each State/Territory
3.17 Highest Railway Station in each State/Territory
3.18 Australian Volcanoes
3.19 Some Large Australian Earthquakes
3.20 Earthquake Hazard Potential in Australia

THE LAND: GORGES AND VALLEYS, CAVES AND CRATERS

4.1 Deepest Gorges in each State/Territory
4.2 Depths of some Well-Known Gorges
4.3 Some Large and Deep Valleys
4.4 Greatest Hillslope in each State/Territory
4.5 Some High Inland Cliffs and Near-Vertical Hillslopes
4.6 Longest Cliffs in each State/Territory
4.7 Australia's Spectacular Roadways
4.8 Greatest Road Ascent in each State/Territory
4.9 Greatest Railway Ascent in each State/Territory
4.10 Australia's Longest Non-Metropolitan Railway Tunnels
4.11 Australia's Highest Bridge Clearances
4.12 Some of Australia's Highest Bridge Structures
4.13 Longest Cave in each State/Territory
4.14 Australia's Deepest Caves
4.15 Some Australian Big Holes
4.16 Australian Meteorite Craters

THE WATER: STREAMS, WATERFALLS AND LAKES

5.1 Australia's Longest Rivers
5.2 Longest River in each State/Territory
5.3 Australia's Largest Catchment Areas
5.4 Australia's Longest Bridges
5.5 Longest Road Bridge in each State/Territory
5.6 Longest Railway Bridge in each State/Territory
5.7 Highest Waterfall in each State/Territory
5.8 Heights of some Well-Known Waterfalls
5.9 Australia's Largest Natural Lakes
5.10 Largest Natural Lakes in each State/Territory
5.11 Australia's Highest Natural Lakes
5.12 Highest Natural Lake in each State/Territory
5.13 Some Lake Depths
5.14 Australia's Largest Reservoirs
5.15 Largest Reservoir in each State/Territory
5.16 Australia's Highest Dams
5.17 Highest Dam in each State/Territory
5.18 Australia's Longest Dams
5.19 Some Australian Hot Springs

THE COAST

6.1 Coastline Length for each State/Territory
6.2 Some Australian Gulfs and Embayments: by area
6.3 Australian Capital City Embayments
6.4 Longest Beach in each State/Territory
6.5 Largest Lagoon in each State/Territory
6.6 Highest Sea Cliffs in each State/Territory
6.7 Some High Sea Cliffs and Near Vertical Hillslopes
6.8 Dramatic Coastlines in each State/Territory
6.9 Some Long Australian Jetties
6.10 Coral Reefs
6.11 Number of Australia's Islands
6.12 Australia's Largest Islands
6.13 Largest Island in each State/Territory
6.14 Lengths of some Australian Straits
6.15 Tides
6.16 Spring Tidal Ranges in the Capital Cities

LIST OF TABLES

9

THE BUSH
7.1 Species Numbers of Common Australian Woody Plant Types
7.2 Species Numbers of Common Australian Herbal and Other Plant Types
7.3 Growth Rates of Trees and other Plants
7.4 Ages of Trees and other Plants
7.5 Australia's Tallest Tree Species
7.6 Some Tall Australian Trees
7.7 Some Tall Shrub Species
7.8 Some Australian Wetlands
7.9 Australia's Vegetation Formations: Pre-1788

THE SKY
8.1 Average Daily Sunshine in the Capital Cities
8.2 Rainy Days and Clear Days in the Capital Cities
8.3 Locations Recording the Highest Yearly Average Temperature in each State/Territory
8.4 Hottest Location in each State/Territory during the Hottest Month
8.5 Locations Recording the Lowest Yearly Average Temperature in each State/Territory
8.6 Coldest Location in each State/Territory during the Coldest Month
8.7 Wettest Location in each State/Territory
8.8 Location with the Most Rainy Days in each State/Territory
8.9 Average Annual Rainfall in the Capital Cities
8.10 Driest Location in each State/Territory
8.11 Location with the Least Rainy Days in each State/Territory
8.12 Highest Recorded Temperature in each State/Territory
8.13 Former Official Highest Recorded Temperature in each State/Territory
8.14 Highest Recorded Temperatures in the Capital Cities
8.15 Highest Recorded Minimum Temperatures in each State/Territory
8.16 Highest Recorded Minimum Temperatures in the Capital Cities
8.17 Lowest Recorded Temperature in each State/Territory
8.18 Lowest Recorded Temperatures in the Capital Cities
8.19 Lowest Recorded Maximum Temperatures in each State/Territory
8.20 Lowest Recorded Maximum Temperatures in the Capital Cities
8.21 Some Cold Coastal Temperatures
8.22 Highest Rainfall in One Day, for each State/Territory
8.23 Highest Rainfall in One Day, in the Capital Cities
8.24 Location Recording the Highest Annual Rainfall in each State/Territory
8.25 Some Significant Australian Bushfires
8.26 Some Significant Australian Droughts
8.27 Some Significant Australian Floods
8.28 Wind Speeds and the Strongest Recorded Wind Gusts in the Capital Cities
8.29 Some Significant Australian Storms
8.30 Incidence of Snow: by State/Territory
8.31 Incidence of Snow: by Capital Cities

LIST OF FIGURES
1.1 Comparative Sizes of Built-Up Areas
2.1 Latitudinal Cross-Sections of Australia
2.2 Longitudinal Cross-Sections of Australia
3.1 Comparative Size of the Biggest Mountains in each State/Territory
4.1 Cross-Section of some Valleys and Gorges in Australia
4.2 Cross-Sections of Greatest Hillslopes in each State/Territory
4.3 Cross-Sections of Ranges and Valleys near the Capital Cities
5.1 Comparative Length of Australian Rivers
5.2 Comparative Sizes: The Largest Lakes in each State/Territory

LIST OF SYMBOLS AND ABBREVIATIONS
STATES AND TERRITORIES
ACT	Australian Capital Territory
Aus	Australia as a whole
CSIT	Coral Sea Island Territory
NSW	New South Wales
NT	Northern Territory
OSA	Offshore Australia
Qld	Queensland
SA	South Australia
Tas	Tasmania
Vic	Victoria
WA	Western Australia

LOCATION ON THE EARTH'S SURFACE
°	degrees
′	minutes

LENGTHS, HEIGHTS, DEPTHS, AREAS
ft	
ha	feet
km	hectares
m	kilometres
mm	metres
sq km	millimetres
	square kilometres

TEMPERATURES
°C	
°F	degrees Celsius
	degrees Fahrenheit

OTHER SYMBOLS
APPROX	APPROXIMATE
E	
M	east
na	million
S	not available
	south

CHAPTER 1
DISCOVERY AND SETTLEMENT

CHAPTER 1
DISCOVERY AND SETTLEMENT

"Aborigines believe that, in the beginning there were floods, then a great void, then the creating beings woke out of a deep sleep. These creating beings, the kangaroo, caterpillar, penguin and so on, travelled a special path - creating along the way - then settled down in some ossified form, a river, a mountain, waterfalls, a flow of bushes. These forms are the sacred sites. They are not dead, but still living. Each Aborigine relates to one particular creative being and its sacred site is the on-going force of his life today."

Mick Dobson, The Age, 1981

Aboriginal encampment in SA

Aboriginal Occupation

Aborigines have occupied Australia for a very long time. The entire continent has been occupied by them, including parts of its undersea margins during times of lower sea levels. Their basic unit of occupation, as described by non-Aborigines, was the 'tribe'; but perhaps a better term is 'clan group', as the word tribe suggests strict boundaries. A clan group is drawn together by some common identity. It is a group of people related by mythical and real connections to a piece of country, over which they had the exclusive rights to exist by hunting, fishing, gathering and by the practice of rituals.

At different times various clan groups would congregate for festivals and feasts; other gatherings would occur for trade or ceremonial purposes. Clan groups varied in number from well below 100 people to as many as 400-500, and the size of each group's economic range basically depended on its available food supply. This, in turn, was dependent on rainfall; it is thought that in arid regions clan groups required about 50,000sq km of land and in areas of moderately high rainfall, about 5000 sq km.

Aborigines had, and many still do have a profound relationship to their land. Basic to their beliefs is the idea that they share the same life force with all the elements of the landscape and all the natural species within it. For the Aborigines this view of the life-force is expressed via the concept of the Dreaming.

Aborigines believe that during the Dreaming ancestral beings with supernatural power and energy traversed the Earth and, as they travelled, formed the topography of the land. They believe that the energies of these ancestral beings remained embodied within the earth along the tracks they followed and at special places where special events took place. The places where the beings stopped are the sacred sites. Such sacred sites include hills, rocks, trees, waterholes, and many other natural objects. The significance of the sites is that part of the spiritual substance of life is contained there, so that as Aborigines move through the landscape, the landscape would provide signs of this spiritual substance. The Aborigines gained their energy (their purpose of living) from these sites, the Dreamtime tracks, and the land in between. This is proudly proclaimed as 'my country'. There are many named sites in the Aboriginal landscape. Some are related to plants, animals or natural phenomena

DISCOVERY AND SETTLEMENT

such as wind or lightning: these are called totemic sites, and may or may not be 'sacred' (in the non-Aboriginal sense). Other sites relate to the supply of food (plants and animals) and resources (stones, timber, bark, medicine, tobacco, and so on), the distribution of which controls the traditional patterns of movement.

This connection between the caretakers, the land and the Dreamtime is maintained by ceremonies performed by the clan. Through the ceremonies the Dreamtime energies are kept alive, and the knowledge is passed on to successive generations. Past, present and future is regarded as an uninterrupted cycle. The spirit of the Dreaming, being brought to life by the birth of children, is passed on to the people, the land and all it contains through various rituals. As it is the source of all life, everything the Dreaming touches is sacred, and sacredness becomes a part of living. Initiation practices reinforce the Aborigine to the sacred. Death is another form of initiation, which returns the spirit to the Dreaming; this, turning in full circle, once again creates life from death.

Obviously the land is important to the Aborigines. But if it is taken away from them — if they become dispossessed, it means that the ceremonies break down. The result is that the energies are reduced; the life force is lost and, consequently, so too is the Aborigines' identity. In its place are the ills of all dispossessed people: ill health, drunkenness, violence, and so on. For the Australian Aborigines, to lose the land is to lose themselves.

Recent First Contacts

Some 'first contacts' have occurred in the recent past. The Pintupi, along with the Wangkatja people of the Great Victoria Desert, were the last large groups to be 'brought in' during the clearance of the Woomera Rocket Range in the 1950s and early 1960s. In the late 1960s and early 1970s there were still a number of Budidjara, Gadudjara, Ngdadjara and Wanman people living traditionally in the western arid regions. By the mid 1970s the traditional lands of these

---

The dates in table 1.1 are approximate and indicate the time of likely first contact between Aboriginal groups and Europeans; coastal Ngarrindjeri and Kurnai people probably had earlier contact with Europeans. It is possible that most groups would have had some knowledge of the Europeans before first contact. Contact was not immediately followed by occupation and settlement, particularly in the desert areas, although once the pastoral and agricultural frontier pushed further inland Aboriginal groups faced major disruptions to their lives and, in most cases, loss of their land. Of the groups listed in this table only the Gagudju, Pitjantjatjara, Walpiri and Pintupi people maintained their traditional way of life well into the 20th century.

Table 1.1 First Contacts between Aborigines and Europeans		
Date	Clan Group	Region
1630s	Tiwi	Melville-Bathurst Islands, NT
1640s	Gagudju	Top End, NT
1770s	Eora	Sydney, NSW
1770s	Banjelang	North Coast, NSW
1830s	Narrenjeri	Murray River, SA
1830s	Barkinji	Western Plains, NSW
1840s	Kurnai	Gippsland, Vic
1840s	Dieri	Cooper Creek, SA
1860s	Arrernte	Central Australia, NT
1860s	Mitakoodi	North-West, Qld
1860s	Kalkadoons	North-West, Qld
1870s	Pitjantjatjara	Western Desert, SA-WA-NT
1870s	Walpiri	Tanami Desert, NT
1890s	Pintupi	Great Sandy Desert, WA-NT

Captain Cook

peoples were virtually empty for the first time in possibly 30,000 years. Following this, in 1977 the last two Mandildjara people of the Gibson Desert, Western Australia, were encouraged to join their kinsfolk in the settled districts.

In October 1984 a small group was contacted in the Stansmore Range area of the Great Sandy Desert, Western Australia. Although at the time this was thought to be a first contact, it seems likely that they knew about life in the outside world. The last contact was in 1986 when seven Wangkatja people decided to join their kin at an Aboriginal settlement on the edge of the Great Victoria Desert, Western Australia. During an expedition in the late 1980s there were some Western Desert Aborigines who believed that there were still some small groups living traditionally in those parts. Interestingly, the informants described themselves as 'going to Australia' when leaving their traditional country for the settled districts. Of course, with the granting of land rights, coupled with the homeland or outstation movement, many semi-traditional Aborigines are returning to their own country.

Discovery and Exploration

The Aboriginal presence in Australia can be confidently dated back approximately 60,000 years. As such, they were the first discoverers, explorers and occupants of the Australian landmass.

None of the early European explorers happened upon the more benign east coast of Australia until Cook, although some people believe that the Portuguese may have travelled along Australia's southern and eastern coasts. One piece of inconclusive evidence supports this theory: the supposed remains of a 'mahogany ship', which may have been Portuguese in origin. Last seen in the 1880s, it is now deeply buried in sand dunes west of Warrnambool, Victoria.

Further support for this theory that Australia was known to the Portuguese comes from an early map known as the Dauphin Chart, published

Tiwi

circa 1536. On this map was depicted a landmass bearing some similarity to Australia, which was called *Java La Grande*. It is not beyond the realms of possibility that the Portuguese visited Australia, and that Mendonca, around 1521, may have explored its coastline. The Portuguese settlement in Timor was only some 500 kilometres off the northern Australian coast. Various maps were published in succeeding years, among them the Frobisher Map of 1578 and the Wytfliet Map of 1597, both of which called Australia "Terra Australis".

Any European exploration of Australia may, however, have been preceded by the Chinese. From 1405 to 1453 the Chinese navigator Cheng Ho made several voyages south to the Timor area and could have made contact with Australia. The scant evidence for this includes a Chinese soapstone carving dating from the 1400s, which was found near Darwin, Northern Territory, and a piece of porcelain located in the Gulf of Carpentaria region. Given that the Chinese had visited the east African coast, a voyage to Australian shores is not unrealistic. Other Asian contacts include the Bajo fishermen (known as sea nomads, and currently based on Roti Island, Indonesia), who have been fishing around Ashmore Island for over 400

DID YOU KNOW?
The earliest known European contact with Australia was in 1606 by a visit by Jansz in the Duyfken.

years, as evidenced by groves of coconuts, garden remains and graves. It is said that Marco Polo heard about a land he called Java Major in 1295.

In 1606 the explorer de Quiros thought he had discovered the southern continent when he found what in fact was one of the islands of Vanuatu, which he named La Austrialia del Espiritu Santo. In *circa* 1628 a Dutch map referred to Australia as Terra del Zur (literally, Land of the South) and a later Dutch map, the Janssonius' map of 1657, was one of the first to show the western portion of the Australian coastline virtually complete from Cape York to the Nullarbor Plain, as well as parts of Van Diemens Land. Australia was referred to as Hollandia Nova or New Holland.

In 1770 the eastern half of the Australia continent was claimed by the Englishman Captain Cook, who, it is alleged, named it New South Wales. In fact, it was the ghost-writer of his journals, Dr Hawkesworth, who originated this name. Nowhere in Cook's or his officers' original papers does the name New South Wales occur.

By the time the first settlement was established at Sydney Cove in 1788 most of the Australian coastline was known to the Europeans, the exception being parts of the south and south-east. French explorer D'Entrecasteaux discovered the Derwent River estuary in 1792-93. Bass, in his open whaleboat, sailed the Gippsland coast to Western Port in 1798. In the same year Flinders sailed amongst the Furneaux Group of islands, off Tasmania's north-east coast. Both men concluded that Van Diemens Land was an island and, together, circumnavigated it in 1798-99. Grant sailed through Bass Strait along the Victorian coast in 1800. The French navigator Baudin sailed westwards along the Victorian and South Australian coasts in 1801, meeting Flinders at Encounter Bay. Flinders was heading east after having explored the coast between Nuytsland and Encounter Bay, including the Spencer and St Vincents gulfs.

Dampier    *Flinders*    *Cook*    *La Pèrouse*

Table 1.2 Exploration and Discovery of the Australian Landmass: Before 1788

Date	Explorer(s)	Location Explored/Discovered
60,000BC	Aborigines	Australia's north coast
3,000BC	Melanesians	Torres Strait Islands
?	Malays?	Australia's northern coast?
1400s	Cheng Ho?	Australia's northern coast?
1500s	Portuguese explorers?	Australia's north-west coast?
1503-04	de Gonneville?	Terre Australe?
1521	Mendonca?	Java La Grande (Australia's east coast)?
1600s?	Bajo fishermen	Ashmore Reef
1606	Jansz	Cape Keerweer, Qld
1606	Torres	Torres Strait, Qld
1616	Dirk Hartog	Dirk Hartog Island, WA
1618	van Hillegom	North-West Cape, WA
1618	Jansz	Exmouth Gulf, WA
1619	Edels	Edelsland, WA
1619	Houtman	coast south of Perth, WA
1621	Brooke	Barrow Island, WA
1622	Brooke	Eendracht Bay, WA
1622	Gerritsz	Cape Leeuwin, WA
1623	Carstens	Cape York Peninsula, Qld
1623	van Colster	Arnhem Land, NT
1627	Nuyts	Nuytsland, WA
1627	Thijssen	Nuyts Archipelego, Cape Leeuwin, WA
1628	Witts	Wittsland, WA
1629	Jacobsz	Houtman Abrolhos Islands, WA
1629	Pelsart	Houtman-Abrolhos-North West Cape, WA
1636	Pieterszoon	Van Diemens Gulf-Melville Island, NT
1642	Tasman	Van Diemens Land, Tas
1644	Tasman	Gulf of Carpentaria, NT-Wittsland, WA
1681	Daniel	Houtman Abrolhos Islands, WA
1688	Dampier	Dampier Land, WA
1697	de Vlamingh	Dirk Hartog Island, WA
1699	Dampier	Shark Bay, WA
1700s	Macassan voyagers	Marege (Arnhem Land, NT)
1768	Bougainville	Great Barrier Reef, Qld
1770	Cook	New Holland (Australian east coast)
1772	du Fresne	Van Diemens Land
1772	St Allouarn	Cape Leeuwin-Shark Bay, WA
1777	Furneaux	Van Diemens Land, Tas
1788	La Pèrouse	Botany Bay, NSW

Murray filled in a gap by exploring part of Port Phillip Bay in 1802; his work was completed by Grimes in 1803.

In 1805 Flinders championed the named Australia, which appeared on his 1814 map, the 'General Chart of Terra Australis or Australia'. Governor Macquarie was the first to officially use the name Australia, on a document in 1817.

Early European Settlement

Many of the early European explorers and settlers saw the Australian landscape as an unfriendly place. With the growth and spread of European occupation during the 19th and early 20th centuries, most settlers were concerned with wresting a living from what they saw as a hostile environment. The European settlers came from lands that were closely settled and green, and had weather that was generally cool and moist – even if in their immediate past some were more accustomed to the impoverished dwellings and habitats of industrial society.

Despite Australia being hotter and drier than Europe, it was optimistically thought that settlement would push back the margins of the desert – an idea that faded only one or two generations ago.

In fact the opposite occurred. The animals' tread compacted the soil and the rain ran over the surface, eroding it and infilling the creeks and waterholes with sediment. The plentiful rains of a few successive years that led people further and further towards the margins of the desert were aberrations.

Dry seasons followed with crop failures and sheep deaths. With practices such as the overstocking of sheep on the western plains of New South Wales and the advancing cropping frontier of South Australia, the soil compacted and the vegetation was removed so that the top soil lost its natural protective covering and became prone to wind erosion. The hot north-westerlies of spring carried the by-then powdery dust into Adelaide, Melbourne or across the Tasman Sea to New Zealand. Dust storms became an Australian climatic feature.

The European settlers had little or no experience of climatic fluctuations, particularly rainfall. Believing that 'rain would follow the plough', they continued to march off into the marginal country during the (unknown to them) good seasons. Areas are considered marginal when they are only capable of producing crops now and then, owing to low and unreliable rainfall. Instead of being pushed back, the desert itself pushed towards the settled districts: the desertification of Australia had begun. Even today the erosion scars and depleted rangelands wrought by these activities are still evident. Who can forget that final dramatic scene at the end of the 1982-83 drought when top soil from the marginal, already established Mallee wheatlands was dumped by a giant dust storm on Melbourne, reducing visibility to less than 100m.

As a result of over 200 years of European occupation the face of the Australian countryside has changed. Australia still retains

Captain James Cook

The Endeavour

DISCOVERY AND SETTLEMENT

Table 1.3 First Temporary European Settlement in each State/Territory

NSW	La Pèrouse camp, Botany Bay	1788
Vic	Sullivans Bay convict camp, Mornington Peninsula	1803
Qld	Redcliffe camp, Moreton Bay	1824
SA	Reeves Point (Kingscote), Kangaroo Island	1836
WA	Houtman Abrolhos islands	1629
Tas	Risdon Cove, near Hobart	1803
NT	Fort Dundas, Melville Island	1824
ACT	Limestone Plains (Ginninderra)	1826

Inauguration at La Pérouse

a significant area of natural heritage, albeit somewhat altered, primarily within the arid lands and, to a lesser extent, in pockets along the eastern and Tasmanian ranges. This wilderness and Aboriginal country accounts for approximately 27 per cent of Australia's area. A further 60 per cent of Australia's area has been utilised for the extensive grazing of sheep and cattle. These outback rangelands still retain significant areas of native vegetation, though with varying degrees of degradation. A further small area, around 2 per cent of the landmass, exhibits varying degrees of alteration – these are forested lands utilised for logging and woodchipping.

Of all of Australia's rural land uses that have been most altered are the agricultural and intensive grazing areas. Wholesale land clearances have removed most of the native vegetation, replacing it with agricultural crops and improved pastures. These landscapes, known variously as 'the wheatbelt', 'broad-acre farming lands' and so on, account for about 10 per cent of Australia's land area.

The most significant impact of European occupation and settlement can be found within the 1 per cent of Australia's area devoted to urbanisation, including not only cities and towns but also transport networks, powerlines, mines, industrial areas and ports.

Today, the ecological footprint is 4.4ha of productive land per person in Australia. This is how much space each individual requires to produce food, housing, transport, consumer goods and services in order to maintain our standard of living.

Non-European Settlements

In Victoria there are the remains of a large stone village constructed by Aborigines near Lake Condah, in the Western Districts. In one area there are the ruined walls of up to 175 dwellings capable of housing up to 1000 people. The site is about 2000-3000 years old. In Queensland non-European settlements have existed on the Torres Strait islands for many hundreds of years. The cays of Ashmore Reef, off the north-west coast of Australia, have been visited by Bajo fishermen for centuries, as evidenced by garden remains and graves.

In the Northern Territory are the remains of temporary settlements established by the Macassans in the 18th century. The Macassans had visited northern Australia since the 1700s, perhaps earlier, to fish for beche-de-mer (sea slugs),

In New South Wales, the French navigator La Perouse, having met the First Fleet in Botany Bay on 26 January 1788, set up camp for six weeks on the northern shore of the bay. In Victoria, scant evidence remains of the Sullivans Bay camp, with some references stating that the convict camp at Western Port Bay, established in 1826, is the oldest European settlement. Reference should be made of Captain Cook's lay-over on the Endeavour River at Cooktown, Queensland, in 1770. A temporary camp was established while he had his damaged ship repaired after striking a coral reef. In South Australia, it is likely that sealers' or whalers' camps predate the Reeves Point settlement mentioned in table 1.3. The temporary settlement mentioned in this table for Western Australia was that of the survivors of the *Batavia* wreck, who lived on the Houtman Abrolhos islands and built fortifications there.

A number of attempts were made to establish a European presence in northern Australia. In the Northern Territory, after the failed Fort Dundas settlement, attempts were made in 1827 to establish a settlement at Raffles Bay, on the Coburg Peninsula and, in 1838, at Victoria, Port Essington, also on the Coburg Peninsula. At Booby Island, in Torres Strait, a landing stage, flagstaff and box were constructed in 1836 by Captain William Hobson. The box acted as a 'post office' or message station for passing mariners as well as a letter exchange and forwarding facility.

DID YOU KNOW?

The **oldest continually inhabited place in Australia** is at Malangangerr in the Kakadu National Park, Northern Territory. Until the 1970s it had been occupied for 23,000 years.

"The soils of the plains is loose, and in very dry weather the grass nearly disappears; but as the country becomes stocked and the tread of the animals binds the surface; the grass acquires closeness and strength and the saltbush vies way to the characteristics of the [Western] slopes [of New South Wales]. As a consequence the rain that falls begins to form watercourses, watercourses become creeks and the streams increase in volume."

G.H. Reid, 1874 (who later became Prime Minister of Australia, 1904-05)

View in Sydney Harbour looking west

Table 1.4 Oldest European Settlements in Australia

Sydney, NSW	1788
Kingston, Norfolk Island	1788
Parramatta, NSW	1790
Hobart, Tas	1803
Newcastle, NSW	1804
Launceston, Tas	1804
Georgetown, Tas	1804
New Norfolk, Tas	1807
Liverpool, NSW	1810
The Macquarie Towns, NSW	1810

Table 1.5 Oldest European Settlement in each State/Territory

NSW	Sydney Town	1788
Vic	Portland	1834
Qld	Brisbane	1824
SA	Adelaide	1836
WA	Albany	1826
Tas	Hobart Town	1804
NT	Palmerston (Darwin)	1868
ACT	Tharwa	1862
OSA	Kingston, Norfolk Island	1788

The dates in tables 1.4 and 1.5 are the respective years of establishment for each settlement. Tasmania at this time was, of course, known as Van Diemens Land. The Macquarie Towns included Pitt Town, Windsor, Wilberforce, Castlereagh and Richmond. Bathurst, on the Central Tablelands of New South Wales, which is considered to be Australia's first 'inland' town, was established in 1815. The first inland city in Australia was Goulburn, New South Wales, proclaimed in 1863.

By way of interest, the oldest known evidence of British colonisation are the initials and date 'FM 1788' (sounds like a radio station) carved in sandstone rock at Garden Island, Sydney by Frederick Meredith from the First Fleet ship, the *Sirius*.

Scene of Sydney from Lavender Bay

View of Sydney from the government paddock

which they transhipped, after curing, to China. They sailed south-east each summer monsoon from the Celebes in fleets of up to 30 or more praus. In Australia they established settlements of platform houses on stilts, with steeply pointed roofs of coconut leaves that they brought with them. These settlements are still marked today with groves of tamarind trees and the ruins of stone fireplaces and pits.

Old Australian Buildings

What is a building? A building consists of surfaces and spaces and all the components therein. A Chinese philosopher, Lao Tzu, once said that "though clay may be moulded into a vase, the essence of the vase is the emptiness within". The same can be said about buildings.

The different functions of designed spaces has lead to different building structures that influence the user. Many large banks and government buildings constructed during the colonial period, for instance, incorporated classical architectural spaces in their design to impress the users of the building. The classical space created relied on a vast scale to overwhelm the user, with power and authority in the case of government buildings, or wealth and security in the case of banks. Church authorities, too, could see the value of space. Many of the great cathedrals used a space that is controlled – on entering a cathedral with a high pitched roof one experiences an upwardly rushed feeling directing one's attention above. In the case of the average suburban house, spaces are tightly controlled, the design limiting that space so that it generally just accommodates a nuclear family, effectively controlling the family size by limiting the number of rooms.

A building's function is closely related to a society's lifestyle. In a western-based society, such as in Australia, these functions can be subdivided into 8 types: residential; commercial; institutional; entertainment; transport; utilitarian; special purpose; landscaped environments.

Residential functions are the most common use of buildings. These may be single dwellings, semi-detached dwellings, terraces, flats, home units, apartments, studios, villas, hotels, motels and temporary structures. Commercial functions include buildings used for shops, shopping malls, warehouses and office blocks including skyscrapers. Institutional functions are housed in such buildings as hospitals, government centres, schools, prisons, churches and museums. Entertainment functions include theatres and cinemas as well as landscaped ground such as racecourses, sporting stadia, drive-ins, etc. Transport functions are specific functions related to the movement of goods and people: buildings include railway stations and related superstructures, airports, wharves and container terminals, road superstructures, bridges, tunnels, aqueducts. Utilitarian functions incorporate those buildings that are designed with a specific manufacturing, storage or extractive industry in mind, such as factories, industrial complexes, mines, power stations and so on. Special purpose buildings – for instance, fortifications, monuments, towers and statues – include those buildings that do not readily fall into the above categories. Finally, landscaped environments are built structures that may not include buildings - these include: parks; gardens; car parks; cemeteries; rubbish tips.

Of course, not all buildings are necessarily used for the function for which they were intended; old buildings may be recycled for other uses, permitting new functions to occur. Redevelopments, too, may retain original facades of buildings in order to maintain the integrity of a streetscape.

The dates in tables 1.8 to 1.17 (except 1.13) generally refers to year of completion and applies to European buildings and structures. The entries in tables 1.8 - 1.17 should be considered provisional.

Elizabeth Farm, Parramatta

Old Government House, Parramatta

Table 1.6 Some of Australia's Oldest Buildings

Elizabeth Farm, Parramatta, NSW	1794
Old Government House, Parramatta, NSW	1799
Governors Dairy Cottage, Parramatta, NSW	?
Commissariat Store, Hobart, Tas	1808
Ingle Hall, Hobart, Tas	1814
Cadmans Cottage, Sydney, NSW	1816
Entally House, Hapsden, Tas	1819

Elizabeth Farm, Parramatta

Table 1.7 Oldest Building in each State/Territory		
NSW	Elizabeth Farm, Parramatta	1794
Vic	Captain Mills Cottage, Port Fairy	1837
Qld	The Windmill, Brisbane	1828
SA	Holy Trinity Church, Adelaide	1838
WA	The Round House, Fremantle	1831
Tas	Commissariat Store, Hobart	1808
NT	The Residency, Darwin	1871
ACT	Duntroon House, Canberra	1833
OSA	Commissariat Store, Kingston, Norfolk Island	1825

The date for The Residency, Darwin, in table 1.7, refers only to a portion of that building. The Commissariat Store, Norfolk Island, is now known as The Beach Store.

Some of the buildings listed above are predated by a number of ruins: the foundations and drains of the first Government House, Sydney, New South Wales, built in 1788; the probable footings of the Old Government House, Western Port, Victoria, built in 1828; ruins on West Wallabi Island, Western Australia, built by survivors of the *Batavia* wreck in 1629; garden remains at Recherche Bay, Tasmania, constructed by French expeditioners, led by Bruny D'Entrecasteaux, in 1792; the Armoury ruins at Fort Dundas, Melville Island, Northern Territory, built in 1824.

Table 1.8 Oldest House in each State/Territory		
NSW	Cadmans Cottage, Sydney	1816
Vic	Captain Mills Cottage, Port Fairy	1837
Qld	Newstead House, Brisbane	1846
SA	The Grange, Adelaide	1840
WA	Patrick Taylor Cottage, Albany	1832
Tas	Ingle Hall, Hobart	1814
NT	The Residency, Darwin	1871
ACT	The Oaks, near Queanbeyan (NSW)	1837

Captain Mills Cottage, Port Fairy

In table 1.8 the date of completion of Captain Mills Cottage is beaten by two other buildings – the reconstructed Captains Cook Cottage, Fitzroy Gardens, Melbourne, originally constructed in England in 1755 or earlier and deconstructed Lonsdale Cottage, originally fabricated in Sydney in 1836, and shipped to Melbourne in 1837. Originally situated in Jolimont, the cottage was moved to Carrum, Victoria, in 1891. In 1962 the National Trust of Victoria bought the cottage and rebuilt it, bar its shingle roof, at a site in Mt Waverley, Victoria. It was reported recently that the site had been sold, the cottage dismantled and its parts stored at Como House, South Yarra, Victoria.

Table 1.9 Oldest Farmhouse or Homestead in each State/Territory		
NSW	Elizabeth Farm, Parramatta	1794
Vic	Emu Bottom, Sunbury	1836
Qld	Cressbrook, Toogoolawah	1843
SA	Springfield, Williamstown	late 1830s
WA	Strawberry Hill Farm, Albany	1831
Tas	Woolmers, Longford	1818
NT	Owen Springs, west of Alice Springs	1874
ACT	Duntroon Dairy, Canberra	1840s

DISCOVERY AND SETTLEMENT

The Round House, Fremantle, WA

Table 1.10 Oldest Public or Utilitarian Building in each State/Territory

NSW	Old Government House, Parramatta	1799
Vic	Customs House, Melbourne	1841
Qld	Old Government Stores, Brisbane	1828
SA	Government House, Adelaide	1840
WA	The Round House, Fremantle	1831
Tas	Commissariat Store, Hobart	1808
NT	Railway Station, Pine Creek	1888
ACT	Government House, Canberra	1891

Table 1.11 Oldest Church in each State/Territory

NSW	Scottish Presbyterian Church, Ebenezer	1809
Vic	St James Old Cathedral, Melbourne	1842
Qld	St Stephens, Brisbane	1850
SA	Holy Trinity Church, Adelaide	1838
WA	All Saints Church, Upper Swan	1841
Tas	St Mathews, New Norfolk	1823
NT	Wesleyan Methodist Church, Darwin	1897
ACT	St Johns, Canberra	1848

Of the churches listed in table 1.11, St James Old Cathedral, Melbourne, was moved to its present site in 1913; a dwelling built at Australind, Western Australia, in 1840 was bought by the Anglican Church in 1915 to become the Church of St Nicholas; the date for St Johns, Canberra, is the date of the church's consecration.

Of the buildings listed in table 1.10 some doubt exists whether Customs House, in Melbourne, is actually the oldest public building. Only the eastern part of the Government House in Adelaide is dated back to 1840. The Northern Territory entry is provisional.

Controversy surrounds Old Government House, Parramatta, with one school of thought being that the original ground floor was built on orders of Governor Phillip in 1790, the additional floor being added by Governor Hunter in 1799.

Church of St Nicholas, Australind, WA

Table 1.12 Oldest Hotel Building in each State/Territory

NSW	Macquarie Arms Hotel, Windsor	1815
Vic	Port Albert Hotel, Port Albert	1842
Qld	Royal Bulls Head Inn, Drayton	1859
SA	The Mountain Hut, Glen Osmond	1845
WA	United Services Hotel, Perth	1840
Tas	Bush Inn, New Norfolk	1814
NT	Pine Creek Hotel, Pine Creek	1890
ACT	The Oaks, near Queanbeyan (NSW)	1838

1.13 Some Old Continuous Hotel Licences in Australia

Woolpack Hotel, Parramatta, NSW	1800
Hope and Anchor Tavern, Hobart	1807
Launceston Hotel, Launceston, Tas	1814
Bush Inn, New Norfolk, Tas	1825
Surveyor-General Hotel, Berrima, NSW	1835
George IV Inn, Picton, NSW	1839
Lord Nelson Hotel, Sydney, NSW	1842
Crown Hotel, Buninyong, Vic	1842
Caledonian Inn, Port Fairy, Vic	1844
Hero of Waterloo Hotel, Sydney, NSW	1844
Old Spot Hotel, Salisbury, SA	1849

Table 1.13 refers to continuous licences, which is not the same thing as the oldest hotel building. The first 3 continuous licence entries do not exit in their original buildings.

Table 1.14 Australia's Oldest Bridges

Macquarie Culvert, Sydney	1816
Richmond Bridge, Richmond, Tas	1825
Clares Bridge, north of Wisemans Ferry, NSW	1825
old stone culverts, north of Wisemans Ferry, NSW	1829-32
two stone bridges, north of Wisemans Ferry, NSW	1830-31
Lennox Bridge, Lapstone, NSW	1833
Kerry Lodge Bridge, Strathroy, Tas	1835
Ross Bridge, Ross, Tas	1836
Lansdowne Bridge, Cabramatta, NSW	1836
Tacky Creek Bridge, Ross, Tas	1836
Restdown Bridge, Risdon, Tas	1838
Campbell Town Bridge, Campbell Town, Tas	1838
Lennox Bridge, Parramatta, NSW	1839
Towrang Creek Viaduct, near Goulburn, NSW	1839
Perth Bridge, Perth, Tas	1839
Jordon River Culvert, Lovely Banks, Tas	1840
Kempton Bridge, near Melton Mowbray, Tas	1840

DID YOU KNOW?

This list includes all bridge types including culverts and viaducts. Though the Richmond Bridge, in Tasmania, is considered **Australia's oldest bridge**, it is predated by the Macquarie Culvert, located in the Botanic Gardens in Sydney. The bridges and culverts listed above as being north of Wisemans Ferry, New South Wales, are located along the disused Great Northern Road. The Lennox Bridge, near Lapstone, is normally considered to be the **oldest bridge on the Australian mainland**.

Macquarie Arms Hotel, Windsor

The buildings listed in table 1.12 are not necessarily used as hotels today. In Fitzroy, Victoria, the Devonshire Arms Hotel was built in 1842. The listing for Queensland may be unreliable. In South Australia, the Crafers Hotel, at Crafers, has been extensively rebuilt but its central section dates back to the 1840s. In Western Australia a military depot at Mahogany Creek, built in 1839, was extended and converted to become the Mahogany Inn in 1848-49.

Surveyor General Hotel

DID YOU KNOW?

The Bush Inn, New Norfolk, could be considered **Australia's oldest hotel**, having been continually licensed since 1825. The Surveyor-General Hotel, Berrima, is the **oldest hotel on the Australian mainland**. The George IV Inn, Picton, was constructed in 1819. Australia's oldest wattle-and-daub hotel is the Tuena Hotel, at Tuena, New South Wales, built in 1866.

Towrang Creek Viaduct, N

DISCOVERY AND SETTLEMENT

Table 1.15 Oldest Road Bridge in each State/Territory

NSW	Macquarie Culvert, Sydney	1816
Vic	Mustoms Creek Bridge, Caramut	1859
Qld	Lamington Bridge, Maryborough	1896
SA	Brownhill Creek Bridge, Adelaide	1861
WA	McCartney Street Bridge, Greenough	1864
Tas	Richmond Bridge, Richmond	1825
NT	unknown	
ACT	Mill Creek Bridge	1883

Table 1.15 should not be considered final; doubts exist over the oldest road bridge in Queensland.

Adelaide River railway bridge, NT

1.16 Oldest Railway Bridge in each State/Territory

NSW	Menangle Bridge, Nepean River	1863
Vic	Flinders Street Station Bridge, Yarra River	1870
Qld	Myall Creek Bridge, Dalby	1868
SA	Currency Creek Bridge, near Mt Barker	1866
WA	old log beam bridge, Katanning-Nyabing Line	1912
Tas	Perth Viaduct, Perth/Longford Bridge, Longford	1870
NT	Adelaide River Bridge, Adelaide River	1888
ACT	Jerrabomberra Creek Bridge, near Canberra	1914

Not all railway bridges listed in table 1.16 are currently in use. The first railway bridge constructed was the original Lewisham Viaduct, completed in 1855. The Flinders Street Station Bridge, is predated by the abutments of the Hawthorn Railway Bridge over the Yarra River, built in 1861.

The Myall Creek Bridge, Queensland, is a 'problem bridge' - it is doubtful that any of the original bridge remains. A reference states that the Splitters Creek Bridge, west of Bundaberg, built in 1881, is the oldest railway bridge in Queensland.

In South Australia, one reference mentions that the masonry of a bridge at Kapunda predates the Currency Creek Bridge by 7 years. Furthermore, the Black Rock Bridge, in South Australia's mid-north, includes girders from the Chief Street Bridge, built in 1856, on the Port Adelaide Line.

Adelaide River Bridge, Adelaide River

Table 1.17 Oldest Mines in each State/Territory

NSW	coal mines in the Hunter Valley	1797
Vic	gold mines at Ballarat	1851
Qld	limestone and coal mines, Ipswich	1827
SA	Glen Osmond silver-lead mines	1841
WA	Geraldine lead mines, Northampton	1848
Tas	lime works at Altamount and Granton	early 1820s
NT	Yam Creek mines	1873
ACT	mica mines along Cotter River	closed 1908

All of the dates in table 1.17 are surpassed by the Aboriginal ochre mine at Wilga Mia, west of Cue, Western Australia, which dates over 1000 years before the present.

Wilga Mia mine, WA

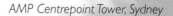

AMP Centrepoint Tower, Sydney

Table 1.18 Australia's Tallest Buildings

AMP Centrepoint Tower, Sydney	309m
Eureka Tower, Melbourne	297m
120 Collins Street, Melbourne	264m
101 Collins Street, Melbourne	260m
Governor Philip Tower, Sydney	254m
Bourke Place, Melbourne	254m
Rialto Towers, Melbourne	251m
Central Park, Perth	249m
Bank West Tower, Perth	247m
Melbourne Central, Melbourne	246m
Chifley Tower, Sydney	244m

The heights in tables 1.18 and 1.19 refer to commercial or residential buildings or towers and are measured to the nearest metre. Excluded are other buildings that could be considered communication towers, chimneys or other structures – see table 1.20 below. Some of these buildings have lightning rods attached, thus increasing their height up to 6m. The Eureka Tower is the tallest residential apartment building in the world.

DISCOVERY AND SETTLEMENT

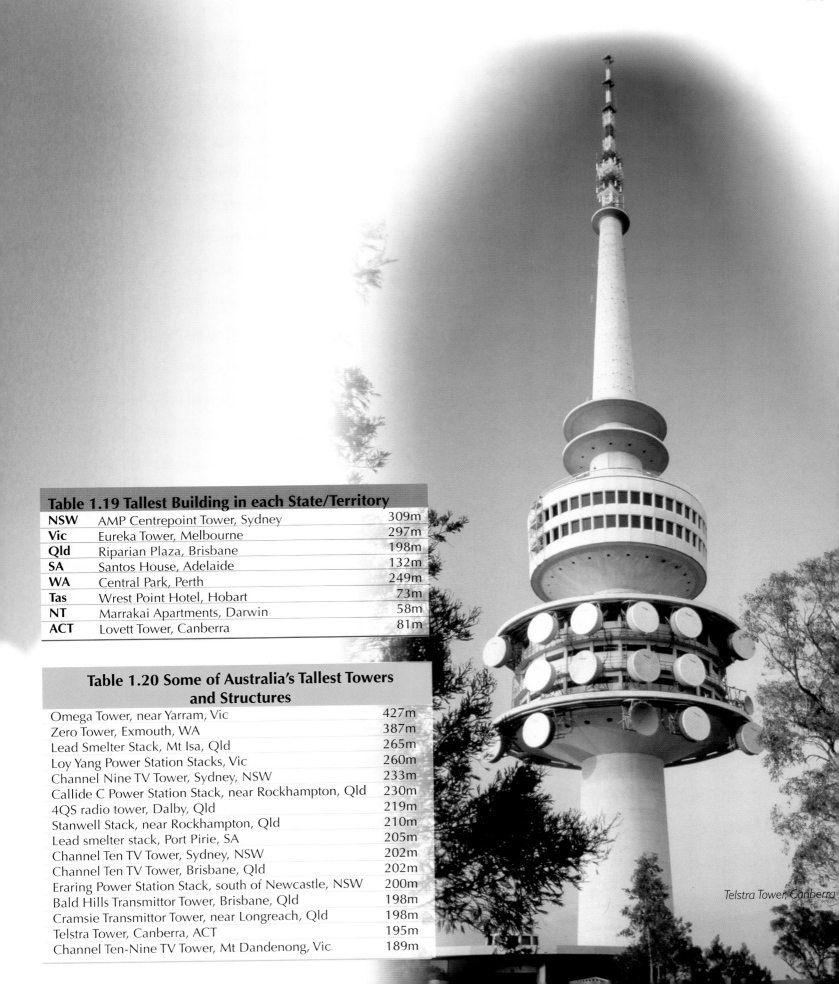

Telstra Tower, Canberra

Table 1.19 Tallest Building in each State/Territory

NSW	AMP Centrepoint Tower, Sydney	309m
Vic	Eureka Tower, Melbourne	297m
Qld	Riparian Plaza, Brisbane	198m
SA	Santos House, Adelaide	132m
WA	Central Park, Perth	249m
Tas	Wrest Point Hotel, Hobart	73m
NT	Marrakai Apartments, Darwin	58m
ACT	Lovett Tower, Canberra	81m

Table 1.20 Some of Australia's Tallest Towers and Structures

Omega Tower, near Yarram, Vic	427m
Zero Tower, Exmouth, WA	387m
Lead Smelter Stack, Mt Isa, Qld	265m
Loy Yang Power Station Stacks, Vic	260m
Channel Nine TV Tower, Sydney, NSW	233m
Callide C Power Station Stack, near Rockhampton, Qld	230m
4QS radio tower, Dalby, Qld	219m
Stanwell Stack, near Rockhampton, Qld	210m
Lead smelter stack, Port Pirie, SA	205m
Channel Ten TV Tower, Sydney, NSW	202m
Channel Ten TV Tower, Brisbane, Qld	202m
Eraring Power Station Stack, south of Newcastle, NSW	200m
Bald Hills Transmittor Tower, Brisbane, Qld	198m
Cramsie Transmittor Tower, near Longreach, Qld	198m
Telstra Tower, Canberra, ACT	195m
Channel Ten-Nine TV Tower, Mt Dandenong, Vic	189m

CHAPTER 2
AUSTRALIA AT LARGE

AUSTRALIA AT LARGE

The Australian Landmass

Australia is a large island situated in the southern hemisphere between the Indian and Pacific Ocean basins. It is a continental landmass whose margins are submerged by relatively shallow waters. The underwater region between the coast and the continental margins is known as the continental shelf. This shelf forms seas, gulfs, bights and straits in some places – for example, the Arafura Sea, Gulf of Carpentaria, Great Australian Bight and Torres Strait. Off the north-eastern coast it also forms extensive barrier reefs, collectively known as the Great Barrier Reef.

The most obvious feature of Australia is that it is a flat and low continent; vast plains cover much of its area. Even the uplands are relatively low tablelands and plateaus, and there are no major mountain ranges. Nonetheless, a geologically unrelated a string of low mountains, plateaus, hills and low rises constitutes the Great Dividing Range. Because of its relative low relief and extensive open areas, the energy systems (stream and wind erosion) operating on Australia's landforms are low. Consequently the land has preserved many ancient geological formations and has the appearance of being vast and old.

Gondwana Legacy

From a comparison of fossil evidence and the shapes of continents at the undersea margins, it is now assumed that the continents were once joined together. Combined with present-day South America, Africa, India, Antarctica, New Guinea and New Zealand, Australia was once a part of the ancient continent of Gondwana.

Australia was still connected with India and Antarctica 150 million years ago when dinosaurs roamed the land. Fossil footprints of dinosaurs can be seen today at Larks Quarry, south-west of Winton, Queensland. As India drifted towards Asia and the Tasman Sea was beginning to open out, shallow seas covered much of the Australian landmass, including the Cretaceous seas that were to become the Great Artesian Basin. Around this time flowering plants appeared. Following the separation of Australia from Antarctica there was considerable volcanic action along the Australian eastern seaboard, and shallow seas covered today's Nullarbor Plain and the Murray Mallee.

Around 55 million years ago Australia's distinctive flora and fauna was evolving independently of the rest of the world. The distinctive shape of Australia's coastline would have been fairly recognisable for the last 50 million years, at least at the edge of the continental shelf. Due to the fact that so little mountain building has occurred since the break-up of Gondwana, Australia has the relatively flat and low-lying landscape we see today.

Over the last one million years geological activity has continually taken place. Climates have changed, and sea levels rose and fell as ice ages came and went. At the last significant lowering of the sea level the Australian mainland was joined with Tasmania and New Guinea, and glaciers covered the high mountains of Tasmania. Minor glaciation also occurred on the high peaks around Mt Kosciuszko. There has been extensive volcanic activity within relatively recent history, resulting in basalt or lava flows in western Victoria and in northern Queensland. Australia's familiar coastline appeared about 6000 years ago, with sea levels then being roughly what they are today. Continental drift has continued throughout this period: Australia is steadily becoming a part of Asia, moving towards it at just less than 5cm a year.

Gondwanaland 300 Million Years Ago

Lizard island, Qld

AUSTRALIA AT LARGE

Table 2.1 Land Area and Population in each State/Territory

NSW	801,600 sq km	6,575,217
Vic	227,600 sq km	4,804,726
Qld	1,727,200 sq km	3,628,946
SA	984,000 sq km	1,511,728
WA	2,525,500 sq km	1,901,159
Tas	67,800 sq km	471,795
NT	1,346,200 sq km	197,768
ACT	2,400 sq km	319,317
CSIT	81 sq km	
OSA		
Australia (total)	7,682,300 sq km	19,413,240

In table 2.1 population figures are based on the 2001 Census; CSIT stands for Coral Sea Islands Territory. It is not included in the total for Australia. The Coral Sea Islands Territory, extending east from the outer edge of the Great Barrier Reef, is Australia's least known territory. Established in 1969, it has a total area of 780,000 sq km – Elizabeth and Middleton Reefs, lying 160km north of Lord Howe Island, were added to the Territory in 1997. The Territory is somewhat unusual in that most of it lies underwater. Composed of the shallow Coral Sea and dotted with numerous coral reefs and sandy cays, it has little land lying above sea level.

Australia is one of the largest countries in the world, ranking sixth in size after Russia (17.1 million sq km), Canada (9.97 million sq km), China (9.59 million sq km), USA (9.36 million sq km), and Brazil (8.51 million sq km).

An idea of the form of the Australian landmass can be seen from the cross-sections shown in figures 2.1 and 2.2.

Table 2.2 Some Facts About Australia

Altitudes

Highest Point
- on Australian territory	Mawson Peak, Heard Island	2744m
- on Australian mainland	Mt Kosciuszko, NSW	2230m
- on Australian island	Mt Ossa, Tas	1617m

Lowest Point
- on Australian mainland	near Silcrete Island, Lake Eyre, SA	-15.2m
Average altitude of Australian land mass		300m
Land over 500m (percentage)		13%
Land over 1000m (percentage)		0.5%

Distances
Distance from Cape York, Qld to South East Cape, Tas	3680km
Distance from Cape York, Qld to Wilsons Promontory, Vic	3180km
Distance from Cape Byron, NSW to Steep Point, WA	4000km

Australia's highest point, Mawson Peak, is the summit of an active volcano, Big Ben, and supports Australia's only glaciers.

DID YOU KNOW?

The Great Dividing Range one of the **longest drainage divides in the world**.

Australia's oldest land surface, found in the Davenport Ranges near the Devils Marbles, Northern Territory, is estimated to be 500 million years old. The **oldest rock crystals in Australia**, found in the Jack Hills, Murchison Goldfields, Western Australia, have been dated at 4300 million years old. In the same region zircon minerals 4100 million years old have been found in rocks at Mt Narryer. Also from the same state are **Australia's oldest fossils**: these are known as stromatolites, discovered at North Pole Well in the Pilbara, and are approximately 3500 million years old.

Cape York, Qld

Table 2.3 Australia's Extreme Geographical Coordinates

Mainland Australia and Tasmania

Most Northerly Point	Cape York, Qld	10° 41'S
Most Southerly Point	South East Cape, Tas	43° 39'S
- on Australian mainland	Wilsons Promontory, Vic	39° 08'S
Most Easterly Point	Cape Byron, NSW	153° 30'E
Most Westerly Point	Steep Point, WA	113° 09'E

Offshore Australia

Most Northerly Point	Bramble Cay, Qld	9° 09'S
Most Southerly Point	Bishop and Clerk Island, Tas	55° 06'S
Most Easterly Point	Norfolk Island	167° 57'E
Most Westerly Point	Cocos Islands	96° 53'E

Claims have been made that Black Head, near Ballina in New South Wales, is mainland Australia's easternmost point at very low tides.

On the west of Australia just a short distance north-north-west of Steep Point, Western Australia, lies the island of Dirk Hartog. The westernmost point of this island lies just south of West Point, at approximately 112° 55'E. The island Pedra Blanca lies roughly 22km south of South East Cape at approximately 43° 51'S.

Table 2.4 Extreme Geographical Coordinates of each State/Territory

New South Wales

North	inland from Point Danger, North Coast	28° 10'S
South	Cape Howe	37° 31'S
East	Cape Byron	153° 39'E
West	western border	141° 00'E
Offshore east	Wolfe Rock, Lord Howe Island Group	159° 08'E

Victoria

North	north-west of Lindsay Point	33° 59'S
South	Wilsons Promontory	39° 08'S
East	Cape Howe	149° 59'E
West	western border	140° 58'E
Offshore south	islands of the Anser Group	39° 12'S

Queensland

North	Cape York	10° 41'S
South	Beardy-Dumaresq River junction	29° 11'S
East	Point Danger	153° 33'E
West	western border	138° 00'E
Offshore north	Bramble Cay	9° 09'S

South Australia

North	northern border	26° 00'S
South	Cape Northumberland	38° 04'S
East	eastern border	141° 00'E
West	western border	129° 00'E

Western Australia

North	Cape Londonderry	13° 44'S
South	Torbay Head	35° 08'S
East	eastern border	129° 00'E
West	Steep Point	113° 09'E
Offshore north	Stewart Island	13° 41'N
Offshore south	South West Island	35° 12'S
Offshore west	Dirk Hartog Island	112° 55'E

Dirk Hartog Island, WA

AUSTRALIA AT LARGE

Tasmania			
North	Woolnorth Point	40° 38'S	
South	South East Cape	43° 39'S	
East	Cape Forestier	148° 22'E	
West	Bluff Hill Point	144° 36'E	
Offshore north	islands of the Hogan Group	39° 12'S	
Offshore south	Bishop and Clerk Island	55° 06'S	
Offshore east	Macquarie Island	159° 40'E	
Offshore west	King Island	143° 50'E	
Northern Territory			
North	Danger Point	11° 07'S	
South	southern border	26° 00'S	
East	eastern border	138° 00'E	
West	western border	129° 00'E	
Offshore north	New Year Island	10° 54'S	
Australian Capital Territory			
North	Hall district	35° 08'S	
South	near Mt Clear	35° 55'S	
East	south of Bungendore (NSW)	149° 24'E	
West	south of Mt Franklin	148° 46'E	
Coral Sea Islands Territory			
North	northern boundary	12° 00'S	
South	southern boundary	24° 00'S	
East	eastern boundary	157° 10'E	
West	near Wreck Bay	approx 144° 00'E	

Coordinates in table 2.4 are given to the nearest minute of latitude or longitude. The latitudes and longitudes for the Australian Capital Territory exclude the Jervis Bay portion of that Territory. Wreck Bay, listed above, lies approximately 60km north-east of Portland Roads, Queensland.

The observant will have noticed that the common borders between Victoria-South Australia and New South Wales-South Australia border are not located along the same longitude. It was intended that both borders would lie along the 141°E meridian. Owing to inaccurate methods used to determine longitude in the 19th century however, the Victoria-South Australia border where they should have met is joined instead at the Murray River, 3km west of the meridian. As records show the border to be inaccurate, South Australia has a claim to a slice of Victoria 3km wide. A peculiar result of this displacement, owing to the meandering nature of the Murray River near the meridian, is that it is possible to stand in Victoria and look east into South Australia!

Similar problems occurred with the Queensland-Northern Territory border when surveyed in the 19th century. Here, the border lies approximately 1km to the west of the 138°E meridian. The Northern Territory once made a claim for this land but it is generally accepted that boundaries are recognised by practice rather than by latitudes and longitudes.

It is just as well that such amicable dealings are made between State governments: otherwise, New South Wales and Victoria would be in trouble. The boundary between these two States mostly runs along the southern bank of the Murray River, but it has sometimes happened that the territory of one State lies on the opposite bank of the river. During times of flood the river sometimes changes course, abandoning some old horseshoe-shaped meander loops and creating new channels. When this occurs, the abandoned loop and adjacent land lie on the opposite bank. Under common law such land is deemed to continue to be a part of the State whose land it was originally. For example, Ouranie Island is a small section of Victoria north of the river, and the locality of Talmalmo is a part of New South Wales situated in Victoria. The next part of Victoria expected to 'cross the border' (it may have already done so) is downstream from Merbein in the Sunraysia district.

Torbay Head, WA

CHAPTER 2

The Human Dimension - A Widespread Settled Population

With a landmass and island distribution as vast as Australia's territory there is a wide distribution of settlements. From north to south it ranges from the near-equatorial regions to cool temperate latitudes; from east to west it crosses eight time zones – four on the mainland, and four on remote islands.

On the Australian mainland and Tasmania there are three main time zones: eastern standard (10 hours before Greenwich Mean Time); Central (+9.5hrs GMT); Western (+8hrs GMT); and a minor time zone: Central-Western (+8.75hrs GMT) – it operates along the Eyre Highway between Caiguna and Travellers Village. Offshore time zones include: Cocos Islands (+8.5hrs GMT); Christmas Island (+9hrs GMT); Lord Howe Island (+11hrs GMT); Norfolk Island (+11.5hrs GMT).

Table 2.5 Australia's Extreme Sitings of Settlement

Most northerly
- settlement — Koedal Boepur, Boigu Island, Qld — 9° 05'S
- town-1 — Christmas Island — 10° 25'S
- town-2 — Thursday Island, Qld — 10° 34'S
- mainland settlement — Seisa, Qld — 10(51'S
- mainland town — Nhulunbuy, NT — 12° 18'S

Most southerly
- settlement — Cockle Creek, Tas — 43° 35'S
- town — Southport, Tas — 43° 25'S
- mainland settlement — Tidal River, Vic — 39° 02'S
- mainland township — Sandy Point, Vic — 38° 50'S
- mainland town — Apollo Bay, Vic — 38° 46'S

Most easterly
- settlement — Kingston, Norfolk Island — 166° 30'E
- mainland town — Byron Bay, NSW — 153° 40'E

Most westerly
- settlement — Home Island, Cocos Islands — 96° 04'E
- mainland settlement — Useless Loop, WA — 113° 22'E
- mainland town — Denham, WA — 113° 32'E

It is possible that the fishing shack settlement at Quobba Point, Western Australia, is marginally further west than Useless Loop.

Lighthouse, Byron Bay

AUSTRALIA AT LARGE

2.6 Extreme Sitings of Settlements for States/Territories

Lord Howe Island

New South Wales
North	Tweed Heads	28° 11'S
South	Wonboyn	37° 15'S
East	Byron Bay	153° 40'S
West	Burns	141° 00'E
Offshore east	Lord Howe Island	159° 04'E

Victoria
North	Yelta	34° 08'S
South	Tidal River	39° 02'S
East	Mallacoota	149° 45'E
West	Serviceton	140° 59'E

Mallacoota

Queensland
North	Seisa	10° 51'S
South	Wallangarra	28° 56'S
East	Coolangatta	153° 32'E
West	Camooweal	138° 07'E
Offshore north	Koedal Boepur, Boigu Island	9° 05'S

Coolangatta

South Australia
North	Pipalyatjara	26° 11'S
South	Port Macdonnell	38° 03'S
East	Cockburn	141° 00'E
West	Travellers Village	129° 00'E

Port Macdonnell

Western Australia
North	Kalumburu	14° 18'S
South	Albany	35° 02'S
East	Wingelinna	128° 58'E
West	Useless Loop	113° 22'E

Albany

Tasmania
North	Stanley	40° 46'S
South	Cockle Creek	43° 35'S
East	Bicheno	148° 18'E
West	Marrawah	144° 42'E
Offshore north	Currie, King Island	39° 56'S
Offshore west	Currie, King Island	143° 52'E

Stanley

Northern Territory
North	Black Point	11° 09'S
South	Kulgera	25° 51'S
East	Alpurrurulam	137° 50'E
West	Docker River	129° 06'E
Offshore north	Minjilang, Croker Island	11° 09'S

Australian Capital Territory
North	Jervis Bay Village	35° 08'S
South	Williamsdale	35° 34'S
East	Jervis Bay Village	150° 43'E
West	Uriarra Forestry Camp	148° 58'E

Jervis Bay

Off-Shore Australian Territories
North	Christmas Island	10° 25'S
East	Kingston, Norfolk Island	166° 30'E
West	Home Island, Cocos Islands	96° 54'E

Christmas Island

Note that the figures in table 2.6 refer to the coordinates of settlements, not necessarily towns.

Table 2.7 Coordinates of the Capital Cities

City	Latitude	Longitude
Sydney	33° 51′S	151° 13′E
Melbourne	37° 49′S	144° 58′E
Brisbane	27° 28′S	153° 02′E
Adelaide	34° 56′S	138° 35′E
Perth	31° 57′S	115° 51′E
Hobart	42° 53′S	147° 20′E
Darwin	12° 28′S	130° 51′E
Canberra	35° 17′S	149° 08′E

The coordinates for the state capitals are based on the location of their major weather stations.

Parliament House, Canberra

AUSTRALIA AT LARGE

Table 2.8 Australia's Most Inland Settlements

Over 900km inland
Tilmouth Crossing, roadhouse, NT
Papunya, Aboriginal settlement, NT

850-900km inland
Haast Bluff, Aboriginal settlement, NT
Glen Helen, tourist settlement, NT
Alice Springs, regional centre, NT
Kintore, Aboriginal settlement, NT

800-850km inland
Santa Teresa, Aboriginal settlement, NT
Hermannsburg, Aboriginal settlement, NT
Ross River, tourist settlement, NT
Aileron, roadhouse, NT
Patjarr, Aboriginal settlement, WA
Areyonga, Aboriginal settlement, NT
Eromanga, grazing and oil township, Qld

Table 2.8 lists, in a possible order, Australia's most inland settlements. Distances are measured from the sea or major bays, gulfs or estuaries, and should be considered approximate. Eromanga, in Queensland, has claims to be Australia's most inland town – unfortunately, it lies barely 800km inland.

If taken as the greatest distance from the sea, the centre of Australia is not, as many people think, Central Mount Stuart in the Northern Territory, for it lies within 800km of the Gulf of Carpentaria. Nevertheless, the nearby Ti-Tree hotel claims to be Australia's most central pub.

Another measure of Australia's centre is the 'centre of gravity' technique. By apportioning an equal unit of weight to over 50,000 points on the coastline, the 'centre-of-gravity' was found to be a point located at 23°7'S/132°8'E – almost the same as the 'furtherest-point-from-the-sea' technique.

Another measure of Australia's 'centre' is the Lambert Centre (25°37'S/134° 21'E – to the nearest minute) located west of Finke, Northern Territory, approximately 710km inland. This was measured by calculating 24,500 points at the high water mark of the coastline, in a somewhat similar fashion to the 'centre of gravity' technique.

A fourth technique is the 'median point' method, whereby the centre is calculated as the midpoint between the latitudinal and longitudinal extremes of the Australian landmass. By this calculation Australia's median point is located at 25°57'S/133° 13'E – to the nearest minute.

DID YOU KNOW?

Of all the settlements listed in table 2.8, Alice Springs could rightly claim to be **Australia's most inland town.**

Arguments have raged over just where is the exact centre of Australia? This, of course, depends on how the exact centre is measured. Though the Australian landmass is of great extent, no portion of it lies further than 1000km from the sea. An area of the countryside west-north-west of Alice Springs, near Derwent Station, Northern Territory, lies over 900km inland. A point, at 23°2'S/132°10'E, could be considered as the **true centre of Australia** by this 'furtherest point from the sea' technique.

Alice Springs, NT

Table 2.9 Most Inland Settlement in each State/Territory

NSW	Tibooburra, grazing township	510km
Vic	Boundary Bend, farming township	350km
Qld	Eromanga, grazing and oil township	800km
SA	Mt Dare, tourist settlement	670km
WA	Patjarr, Aboriginal settlement	830km
Tas	Bronte Park, tourist settlement	90km
NT	Tilmouth Crossing, roadhouse	910km
ACT	Macgregor, Canberra suburb	110km

The figures in table 2.9 represent shortest distance from the sea, or major inlets, and are approximate.

Family Hotel, Tibooburra, NSW

Table 2.10 Australia's Remotest Settlements or Stations

Kiwirrkura, Great Sandy Desert, WA	697km from Alice Springs, NT
Tjuntjuntjara, Great Victoria Desert, WA	672km from Kalgoorlie, WA
Yurr Yurr, Gibson Desert, WA	625km from Wiluna, WA
Kunawarritji, Great Sandy Desert, WA	590km from Marble Bar, WA
Kulumburu, the Kimberley, WA	562km from Kununurra, WA
Carson River Station, the Kimberley, WA	557km from Kununurra, WA
Ilkulka, Great Victoria Desert, WA	551km from Laverton, WA
Kintore, Western Desert, NT	529km from Alice Springs, NT
Theda Station, the Kimberley, WA	514km from Kununurra, WA

The figures in table 2.10 represent distances via the shortest practicable route to the nearest town or township.

DID YOU KNOW?

Rabbit Flat Roadhouse, in the Tanami Desert, Northern Territory, is often cited as **Australia's remotest roadhouse**. It is located 412km from Halls Creek, Western Australia, and 598km from Alice Springs, Northern Territory. Its remoteness will be surpassed when the Ilkulka Roadhouse (listed above) is fully operational. Located on the Anne Beadell Highway, it lies 551km from Laverton, Western Australia, and 779km from Coober Pedy, South Australia.

2.11 Most Isolated Settlement in each State/Territory

NSW	Wanaaring, grazing township	196km
Vic	Dargo, farming township	80km
Qld	Kowanyama, Aboriginal township	386km
SA	Innamincka, tourist township	352km
WA	Warburton, Aboriginal township	585km
Tas	Strathgordon, hydro township	71km
NT	Nhulunbuy, mining town	452km
ACT	Tharwa, grazing township	5km

The distances in table 2.11 refer to the average distance to the nearest service centre along all trafficable roads – it excludes island settlements.

This table also excludes Aboriginal settlements except where they have 'town' functions – both Kowanyama and Warburton are local shire council headquarters.

Innamincka, SA

AUSTRALIA AT LARGE

2.12 Most Isolated Farm or Station in each State/Territory

NSW	Keewong Station	128km from Cobar
Vic	Shannonvale Station	46km from Omeo
Qld	Inkerman Station	285km from Normanton
SA	Cowarie Station	249km from Marree
WA	Carson River Station	557km from Kununurra
Tas	unknown	
NT	Tanami Downs Station	652km from Alice Springs
ACT	Caloola Farm	22km from Tharwa

The distances in table 2.12 indicate the road distance to the nearest town or township – it excludes roadhouses and Aboriginal settlements.

Cowarie Station, SA

Table 2.13 Australia's Longest Highways

Great Northern Highway	Midland (Perth)-Wyndham, WA	3112km
Stuart Highway	Port Augusta-Darwin, SA-NT	2880km
Princes Highway	Sydney-Adelaide via coast NSW-Vic-SA	1990km
Bruce Highway	Brisbane-Cairns, Qld	1703km
Eyre Highway	Port Augusta-Norseman, SA-WA	1680km
North-West Coastal Highway	Geraldton-west of Port Hedland, WA	1289km
Mitchell Highway	Bathurst-south of Augathella, NSW-Qld	1115km
Newell Highway	Tocumwal-Goondiwindi, NSW-Qld	1060km
Barrier Highway	Nyngan-Gawler, NSW-SA	1046km
Landsborough Highway	Morven-east of Cloncurry, Qld	1010km
Pacific Highway	Sydney-Brisbane, NSW-Qld	960km
Sturt Highway	east of Wagga-Adelaide, NSW-SA	949km
New England Highway	Hexham-Yarraman, NSW-Qld	894km
Hume Highway	Sydney-Melbourne, NSW-Vic	880km
Goldfields Highway	Meekatharra-west of Kambalda, WA	801km
Flinders Highway	Townsville-Cloncurry, Qld	776km
Barkly Highway	Cloncurry-Three Ways, Qld-NT	760km
Warrego Highway	Brisbane-Charleville, Qld	754km
Carnarvon Highway	Moree-Rolleston, NSW-Qld	643km
Great Eastern Highway	Perth-Kalgoorlie, WA	594km

These are the longest highways of Australia by name. Distances will include new freeway sections that are not officially part of that highway but are definitely part of that route. In Victoria the Princes Highway is sometimes subdivided into Princes Highway East and Princes Highway West.

The proposed Outback Highway, running between Laverton in WA and Winton in Qld, will be 2562km long, making it Australia's third longest highway. It will be comprised of the Great Central Road, Tjukururu Road, Lasseter Highway, Stuart Highway, Plenty Highway, Donahue Highway and the Min Min Byway – it will also share a common route with the Stuart Highway for 267km.

DID YOU KNOW?
Australia's longest highway by number, Highway 1, circumnavigates most of the Australian coastline: total distance approximately 12,800km.

Great Northern Highway

Table 2.14 Some Long Walking Trails

Bicentennial National Trail	Cooktown, Qld-NSW-Healesville, Vic	5330km
Heysen Trail	Cape Jervis-Parachilna Gorge, SA	1200km
Bibbulmun Track	Kalamunda-Albany, WA	964km
Australian Alps Walking Track	Walhalla-Tharwa, Vic-NSW-ACT	650km
Hume and Hovell Track	Yass-Albury, NSW	440km
Larapinta Trail	Alice Springs-Mt Sonder, NT	223km
Overland Track	Cradle Mountain-Lake St Clair, Tas	80km

The Bicentennial National Trail and Mawson Trail are also used by horse and mountain bike riders.

Australia's Plains

Australia is characterised by its plains. A plain is an extensive area of level to slightly undulating country with perhaps only the odd residual hill or rocky outcrop breaking its uniform skyline. The are a number of different types of plains.

Floodplains are plains bordering a river and are formed by deposits of sediment carried by that river. During a flood the river rises above its banks, covering the plain with water and sediments. Floodplains exhibit a number of landforms such as stream meanders, billabongs and levee banks. Deltas are a special type of floodplain formed where a river enters a lake or the sea – for example, the Burdekin River delta in north Queensland. Small floodplains in hilly country are sometimes known as river flats. There are numerous examples of floodplains throughout Australia, with perhaps the biggest floodplains being found along many of the inland rivers – the Barwon and Gwydir rivers of northern New South Wales, for example, or Cooper Creek below Windorah in Queensland's Channel Country.

Riverine or alluvial plains are plains of alluvium (eroded sediments) deposited by rivers either currently (hence they are floodplains) or in the past. Riverine plains tend to be relatively featureless – just immense expanses of dead-level uniformity interrupted at intervals by shallow, but normally dry watercourses. Much of the Riverina region of New South Wales and the adjacent districts of north-central Victoria are riverine plains.

Peneplains are formed by rivers and rain gradually wearing down or eroding the landscape to such an extent that only very resistant rocky outcrops or low hills stand above the plain. Peneplains tend to gradually slope towards stream channels, along which there will be a floodplain. Most of the Lake Eyre basin is a vast peneplain.

A pediplain is a type of plain found on the edges of uplifted mountain ranges. A pediplain slopes away from the range towards the lowlands and there is often a distinct angle of slope between the pediplain and the range. Streams issuing from the mountains may cut into the pediplain or deposit sediment (alluvium) over the plain in the shape of a fan. These are alluvial fans, and are a type of floodplain. Good examples of pediplains can be seen along the western side of the Flinders Ranges in South Australia.

Volcanic plains are formed by vast outpourings of lava, which cover the pre-existing landscape. They are generally undulating and support volcanic features such as old craters, stony rises (solidified lava flows), lava tubes and volcanic lakes. Most of the Western District of Victoria is a volcanic plain, as is the McBride Volcanic Province of the north-east highlands of Queensland.

Sandplains are Australia's most widespread type of plain. Dominated by vast sheets of sand, they exhibit either broad and low sandy rises barely a metre high, such as what is found in the Tanami Desert of the Northern Territory, or sand dunes up to 20m high. The dunes are mostly longitudinal, with margins stabilised by vegetation but with mobile and oscillating crests subjected to movement by winds. Longitudinal dunes may be hundreds of kilometres long. Between each dune are flat areas, or swales, up to a kilometre wide and covered with sand or occasionally stones. Most of Australia's named deserts are dune-covered sandplains – for instance, the Simpson Desert ot the Great Sandy Desert.

Stony plains come in two types: gibbers or sheets of rock. Gibber plains are slightly undulating, and covered with countless billions of small to not so small pebbles. The sizes of the gibbers are the basis for differentiating gibber plains: very small gibbers form buckshot plains; moderate size gibbers form gravel plains; large gibbers form shingle plains. A special type of gibber plain is the quartz blow – gibber-sized quartz fragments scattered over the ground. Gibber plains are found near old, low tablelands (known as 'jump-ups') or as rolling stony downs. In arid areas gibber plains form Australia's most desolate landscapes – for instance, Sturts Stony Desert and parts of the Channel Country of Queensland.

Rock plains occur as expansive sheets of rock, the best known example being the Nullarbor Plain. This ever-so-slightly undulating plain of limestone exhibits minor features such as sinkholes and caves.

Shield plains exhibit both rocky and sandy surfaces. Normally there are two levels of plain separated by a low, rocky cliff less than 15m high, which is called a breakaway. Normally the upper plain is slightly undulating with outcrops of rock, usually granite, with occasional low, rocky rises, while the lower plain is generally flat and sandy. Shield plains are enormous and widespread in the south-western quarter of Western Australia, and are also found in Central Australia.

Clay plains are also quite common in Australia. These slightly undulating plains (sometimes called downs) are formed on special types of black, clayey soils, that exhibit wide and deep cracks when dry or produce thick gooey ooze when wet. Most of the downs of midwestern Queensland and the Barkly Tableland are clay plains.

Lacustrine plains (or lake beds) are large and not so large flat beds of what is normally a dry lake. A variety of surfaces are found on lake beds – salt, clay or gypsum, sometimes with a thin covering of vegetation. On the prevailing downwind side of the lake there may be sand, clay or gypsum dunes (these dunes are called lunettes in southern Australia). Lake Mungo and the Walls of China in western New South Wales are examples of a lacustrine plain and lunette.

Coastal plains are gently sloping plains developed as a result of the shoreline extending at the expense of the sea. Coastal plains are of low relief with perhaps only a few remnants of low, residual hills breaking the skyline. The Pilbara coast of Western Australia has good examples of coastal plains.

Tidal plains are coastal plains subjected to inundation by the sea. They are flat with perhaps some residual relict dunes rising above the general level of the plain, and are coursed by wildly meandering estuaries, normally outlets of major river systems. Tidal plains are common along the southern shoreline of the Gulf of Carpentaria.

DID YOU KNOW?

The Lake Eyre peneplain is considered to be the **world's largest peneplain.**

AUSTRALIA AT LARGE

Table 2.15 Some Australian Plains: by size

Lake Eyre peneplain, Qld-SA-NT	1,170,000 sq km
Nullarbor Plain, SA-WA	270,000 sq km
Barkly Tableland, Qld-NT	240,000 sq km
Gulf Plains, Qld	200,000 sq km
North-West Plains, NSW-Qld	145,000 sq km
Hay Plains, NSW	70,000 sq km
lava plains west of Melbourne, Vic	23,000 sq km
Darling Downs, Qld	19,000 sq km
Mundi Mundi Plain, NSW-SA	4,300 sq km
Willochra Plain, SA	3,000 sq km
Adelaide Plains, SA	1,500 sq km

The areas given in table 2.15 for the plains are approximate, owing to problems in defining boundaries.

Floodlplain, Monkira Station, Qld

Peneplain, Lyons River Valley, WA

Pediplain, Flinders Ranges, SA

Sandplain, near Musgrave Ranges, NT

Clay plain, Antrim Plateau, WA

The Human Dimension - Life on the Plains

The occupation of Australian plains has resulted in the establishment of vast grazing properties. Anna Creek Station, South Australia, at 19000sq km, is **Australia's largest station** (it may be smaller now); other large stations include Victoria River Downs, Northern Territory – 12000sq km, and Commonwealth Hill Station, South Australia – over 10000sq km.

Coupled with station development was the control of rabbits and dingos. Starting from a small population released near Geelong in 1859, rabbits spread across non-tropical Australia, entering southern New South Wales by 1880 and southern Queensland by 1886. At times the rabbit frontier moved across the country at 100km per year. Western Australia attempted to stem the rabbit tide by the construction of a series of fences. Rabbit Proof Fence No.1 ran for 1827km between Starvation Boat Harbour, in the south, to Eighty Mile Beach, in the north. That failed, to be replaced by Rabbit Proof Fence No.2 – 1167km long.

As popular as the dingo is amongst many Australians, it is, along with wild dogs, a killer of sheep. One solution, to protect the flock of south-eastern Australia, was to fence-in the sheep grazing country. This resulted in the construction of the 5309km-long Dingo Fence – the **world's longest fence**. It runs from west of Ceduna, South Australia, to north of Mitchell, Queensland. At one time the fence was much longer – 9600km, when it included the grazing land of mid-western Queensland.

Coupled with the growth of pastoralism and the need to connect the far corners of Australia with a transport network, it is not surprising that, in a country as flat as Australia, both roads and railways would have lengthy straights along many of their routes.

Fence, Muloorina Station, SA

AUSTRALIA AT LARGE

Table 2.16 Straightest Section of Road in each State/Territory

NSW	Mitchell Highway, Nevertire-Nyngan	59km
Vic	Sturt Highway, west of Culluleraine	25km
Qld	information unavailable	
SA	Oodnadatta Track, Warriners-Engenina Creeks	71km
WA	Eyre Highway, Caiguna-east of Balladonia	148km
Tas	Evandale Road, Nile-Evandale	9.5km
NT	Barkly Highway, Frewena-Dalmore Downs Station	63km
ACT	Northbourne Avenue, Canberra	4.1km

The distances in table 2.16 refer to dead-straight roads, those without even the slightest bends. Such roads are found in plains country. There are many roads crossing Australian plains that are nearly dead straight, but most have slight bends at 15-25km intervals in order to help keep the driver alert. For instance, the Mitchell Highway between Nyngan and Bourke, New South Wales, is always shown dead straight on the road maps, but has slight bends at regular intervals. The original figure found for the Victorian section of the Sturt Highway west of Culluleraine was 75km but realignments have shortened this distance.

Determining Queensland's straightest road is tricky: the Landsborough Highway between Barcaldine and Ilfracombe goes close but has slight bends. The figure for South Australia is unreliable. The minor road between the Comet and Indooroopilly Outstations north of Tarcoola may be longer, while the old unsealed route of the Eyre Highway near Koonalda Station on the Nullarbor Plain had a straight about 86km long.

Western Australia is the place for dead-straight roads; the figure mentioned above is probably a world record, while the North-West Coastal Highway between the Murchison River and Carnarvon has a few very long straights. The Eyre Highway figure given in table 2.16 is a world record; as a matter of interest, a track following an old telegraph line east of Caiguna extended this figure by another 68km, giving a total length of 216km.

Table 2.17 Straightest Section of Railway track in each State/Territory

NSW	between Nevertire-Nyngan, Main West Line	64.5km
Vic	between Violet Town-Benalla, Albury Line	21.7km
Qld	between Barcaldine-Longreach, Central Line	76.5km
SA	between Ooldea-WA Border, Trans-Continental Line	254km
WA	between SA Border-Nurina, Trans-Continental Line	223km
Tas	west of Launceston, Western Line	6.6km
NT	n.a.	
ACT	between Queanbeyan-Canberra, Canberra Line	1km

Before falling into disuse the Main West Line between Nyngan and Bourke, New South Wales, was the third longest straight stretch of railway in the world, at 186km. In New South Wales there are other long straight stretches: on the Broken Hill Line there are three straights each over 70km long, while between Narromine and Bourke, Main West Line, there were only five curves in 327km.

The figure given in table 2.14 for the Australian Capital Territory's straightest section of railway is exceeded by a stretch 4.3km long on the old Bombala Line between Royalla and Williamsdale, where the line forms part of the common border with New South Wales.

DID YOU KNOW?

The straights on the Trans-Continental Line mentioned above have a combined total of 477km, making this the **world's straightest section of railway line.**

Australia's straightest railway line, Cook, SA

Table 2.18 Some Long Australian Pipelines

Dampier-Bunbury pipeline, WA	1541km
Dampier-Kambalda pipeline, WA	1380km
Amadeus Basin-Darwin pipeline, NT	1512km
Moomba-Sydney pipeline, SA-NSW	1300km
Ballera-Mt Isa pipeline, Qld	810km
Ballera-Wallumbilla pipeline, Qld	756km
Sale-Sydney pipeline, Vic-NSW	730km

The pipelines listed in table 2.15 are natural gas pipelines. Unlike oil, pipelines best deliver natural gas direct to the customer for it is impractical to store it in any large quantities.

Bunbury, WA

Australian Deserts

Another distinguishing feature of Australia is its deserts. Defining a desert is difficult, however: if based on rainfall (average falls totalling less than 250mm per year is the normal measurement), then about half of Australia is desert. If, on the other hand, the definition is based on a virtual absence of vegetation, then hardly any of Australia is desert.

Australian deserts are better described as arid lands. What characterises them is low and erratic rainfall, a relatively sparse covering of vegetation in many but not all areas, and countryside deemed unsuitable for the grazing of stock because of unpalatable spinifex grasses or extensive dune systems. Most named Australian deserts are sand dune-covered plains – those distinguished by longitudinal sand dunes with flat areas (swales) inbetween - for instance, the Great Victoria Desert. Some other named deserts are stony deserts: for example, the Gibson Desert.

Table 2.19 Australian Deserts: by size

Great Victoria Desert, SA-WA	longitudinal dunes	348,750sq km
Great Sandy Desert, WA-NT	longitudinal dunes	267,250sq km
Tanami Desert, WA-NT	sandplain	184,500sq km
Simpson Desert, Qld-SA-NT	longitudinal dunes	176,500sq km
Gibson Desert, WA	stony	156,000sq km
Little Sandy Desert, WA	longitudinal dunes	111,500sq km
Strzelecki Desert, NSW-SA	longitudinal dunes	80,250sq km
Sturts Stony Desert, Qld-SA	stony	29,750sq km
Tirari Desert, SA	sand dunes	15,250sq km
Pedirka, SA	stony	1,250sq km

Owing to problems in defining boundaries, the area of each desert included in table 2.19 is approximate, being only indicative of its size. For instance, one source gives the area of the Great Victoria Desert as 420,000sq km. Furthermore, many believe the Sturts Stony Desert is indistinguishable from the stony plains that comprise much of the arid regions of Queensland's Channel Country. Not included above is the Alberga Dunefield in northern South Australia and the unnamed sandplain desert in the Northern Territory south of the Barkly Tableland.

Simpson Desert, Qld

AUSTRALIA AT LARGE

Other Deserts

Australia has other named deserts with no specific geographic location, such as the Western Desert and the Central Desert. The Western Desert is sometimes used to describe the Great Sandy, Gibson and Great Victoria Deserts combined, or the sandy desert country south-west and west of Alice Springs, Northern Territory, centred around Lake Amadeus, Uluru and the Petermann-Rawlinson Ranges. The Central Desert refers to the arid country north of Alice Springs and may include parts of the Tanami Desert and the sandplains along the Sandover River.

Other named deserts which have occasionally been described include the Meekatharra Desert, covering the stony arid lands to the north-west of Meekatharra, Western Australia; the Canning Desert, which is another name for the Great Sandy Desert; the Arunta Desert, best known as the Simpson Desert.

There are also other tracts of land with the name 'desert'. The Little Desert (1,300sq km) in Victoria; nearby, also in Victoria, there is the Big Desert, which adjoins the Ninety Mile Desert in South Australia (combined area: 7,000sq km) – all these areas are uncleared mallee eucalypt country with sand dunes. In South Australia, north-east of the Flinders Ranges near Lake Blanche, is the Cobbler Desert, a small desert of clay knoblets and shifting dunes formed by the voracious appetite of rabbits devouring the vegetation. In Western Australia, the Pinnacles Desert north of Perth is an area of coastal dunes and sandplains that exhibit an array of limestone pinnacles.

Elsewhere are other 'deserts': the so-called 'desert uplands' of spinifex and stunted eucalypts lying just to the east of the road between Aramac and Torrens Creek in mid-west Queensland. There is also the so-called 'wet desert', an area of open heaths on Cape York Peninsula. Nearby, north-west of Laura in Cape York Peninsula is an area simply called 'The Desert'. All of these areas are not strictly deserts, in terms of aridity, but were so-named by early pastoralists as tracts of country unsuitable for the grazing of stock.

The Human Dimension - Crossing the Desert

Australia's vast size means that there are a number of very long and isolated roads and tracks across its deserts, and other areas of the outback.

Table 2.20 Australia's Desert Tracks

Canning Stock Route	Wiluna-Halls Creek, WA	1820km
Wapet Road-Gary Junction Road	Great Northern Hwy-Papunya, WA-NT	1438km
Anne Beadell Highway	Mabel Creek-Yarmarna stations, SA-WA	1136km
Connie Sue Highway	Warburton-Rawlinna, WA	542km
French Line-QAA Line	Dalhousie Springs-Birdsville, SA	440km
Gunbarrel-Heather Highways	Carnegie Station-Warburton, WA	485km
Rig Road-K1 Line-Warburton Track	French Line-Birdsville Track, SA	419km
Wiso Road	Warrego-Lajamanu, NT	392km

Table 2.21 Some Famous Outback Roads

Tanami Track	Alice Springs-Halls Creek, NT-WA (20)	1012km
Gulf Track	Burketown-Roper Bar, Qld-NT (10)	873km
Plenty-Donahue Highways	Alice Springs-Boulia, NT-Qld (25)	821km
Sandover Highway	Alice Springs-Camooweal, NT-Qld (5)	813km
Buchanan Highway	Dunmurra-Halls Creek, NT-WA (10)	767km
Gibb River Road	Derby-Wyndham, WA (45)	715km
Oodnadatta Track	Marree-Marla, SA (25)	614km
Duncan Road	Halls Creek-Timber Creek, WA-NT (5)	610km
Birdsville Track	Marree-Birdsville, SA-Qld (20)	593km
Warburton Road	Laverton-Warburton, WA-NT (10)	565km
Strzelecki Track	Lyndhurst-Innamincka, SA (45)	467km

The Warburton Road (table 2.21) is also known as the Great Central Road, this name extending the route between Laverton and the WA-NT Border - a total distance of 892km.

Do not be fooled by the name 'highway' or 'road' for some of these routes: these are mostly unmaintained desert tracks. Australia's deserts are remote places and intending travellers on these routes should note that crossings should not be undertaken lightly. In some instances convoy travel with an experienced leader is required, as well as permits. All routes are strictly 4WD.

The original Gunbarrel Highway, crossing over parts of which are restricted today, ran between Victory Downs Station, Northern Territory, and Carnegie Station, Western Australia – a distance of 1351km.

Table 2.21 lists Australia's famous desert and other outback tracks, most of which have mostly been upgraded into two lane, gravel and earth outback main roads. All these roads are suitable for travel in conventional vehicles during dry conditions, with care, and some outback driving experience. Distances are measured between settlements at the end of or beyond the named route, and the table also shows the average number of vehicles per day using these routes.

Sandover Highway, NT

CHAPTER THREE
THE LAND: HILLS, MOUNTAINS AND TABLELANDS

THE LAND: HILLS, MOUNTAINS AND TABLELANDS

The land is a complex of phenomena – a rich tapestry of rocks, soils and plant life, which assumes different forms at different places for different reasons. The major observable fact about the land is the way it lies. It forms a continuous surface, virtually level in some places and moderately or steeply sloping at others. The surface is dissected by stream channels and its edges are delineated by coastlines. As an integral part of this surface, there are numerous landforms, the origin of which has involved a number of interacting elements.

Creation of Landscapes

Physical landscapes are controlled by their underlying structure or geology. The geology of a landscape varies from place to place. In some places it is completely exposed, especially in arid areas where there is minimal vegetation cover; at other places it is buried by deposits of previously transported and eroded rock materials. This erosion, transportation and deposition of materials is what creates landscapes. The main agents of erosion in Australia are water and wind. Ice in the form of glaciers has also played a part in sculpturing Australian landscapes, but only a small one (mainly in Tasmania). Since European occupation, land clearances have accelerated the erosive processes because, after clearing, the land itself is left exposed to the agents of erosion.

Rock materials themselves are broken through a series of complex processes known, collectively, as 'weathering'. Weathering is the process in which the atmosphere interacts with rock materials. One result of weathering is rock disintegration: larger rock particles are broken into smaller rock particles, which are then capable of being transported by water or wind (and ice, too). Weathering may also result in rock alteration, whereby rock materials – in conjunction with other elements such as climatic or vegetative influences – produce soils.

The net result of weathering and erosion of the land by natural processes that have been occurring virtually uninterrupted for extended periods of time is that the surface of the land becomes lower in altitude. Thus it can be seen that the appearance of the physical landscape is dependent on many elements. The physical landscape of today is at the same time an original surface (for the processes of natural erosion and deposition are currently occurring) and a relict or historical surface, which is the result of previous interacting natural processes.

Mountain Building

The land can rise, although this process is usually very slow. The earth's crust – the uppermost level, upon which we live – extends downwards about eight to ten kilometres and is in a state of flux. It moves, and as it makes major or minor adjustments a number of geological or structural forms arise. These structures – which include, for instance, folding and faulting – ultimately become the dominating controls of the physical landscape. Folding is where rock layers are thrust up in wave-like forms. Faulting is where layers of rock are fractured.

Excellent examples of folding can be seen east of Alice Springs in the MacDonnell Ranges, along Bitter Springs Creek. A good example of a fault lies to the immediate west of Lake George along the Goulburn-Canberra highway. Many of Australia's mountain ranges are the result of such folding and faulting – for instance the Flinders Ranges, South Australia; the Stirling Ranges, Western Australia; the MacDonnell Ranges, Northern Territory. Different types of mountains occur elsewhere.

Tablelands

Most of Australia's mountain uplands are tablelands (or plateaus – both words mean the same) and are distinguished virtually level to moderating undulating skylines, often bounded by steep, sometimes vertical escarpments. The escarpment edges may be coursed by streams, which create either waterfalls or, if incised into the tableland, gorges and canyons. Good examples of these features can be seen in the Blue Mountains, New South Wales. Some tableland surfaces are capped with basalt or lava flows, which form relatively higher sum-

The folded Bunkers Range, SA

HILLS, MOUNTAINS AND TABLELANDS

mits – the Atherton Tableland, Queensland, is a good example. Other well-known tablelands include the Hamersley Range, Western Australia, and the Arnhem Land Plateau, Northern Territory. Interestingly, early explorers and pioneers used the name 'range' for these tablelands, as from a distance their escarpments do appear to be ranges.

Other types of tablelands may be formed by large granite intrusions. These originate deep within the earth's crust and are slowly uplifted above the original surface, where they then erode. This erosion process leaves behind high, granite massifs (granite outcrops), such as the Mt Buffalo Plateau, Victoria, or much of the Northern Tablelands of New South Wales. Granite formations are often distinguished by granite boulders (tors), balancing rocks or massive granite domes – for instance, the southern coastline of Western Australia between Albany and east of Esperance. Australia's largest exposed singular granite outcrop is Bald Rock, Northern Tablelands, New South Wales, standing 200m high above the surrounding lowlands.

Australia's largest tableland area is the New England or Northern Tablelands of New South Wales – it extends for 300km north to south and up to 130km east to west at elevations over 915m (3000ft). Other extensive tableland areas (and mountains) over 915m in elevation include the Monaro-Snowy Mountains in New South Wales (160km by 120km), the Central Tablelands of New South Wales (120km by 120km), and the Central Plateau in Tasmania (90km by 40km – this plateau is much larger but elevations are mostly below 915m). Smaller tableland areas above 915m include the following: Barrington Plateau, New South Wales; Mt Buffalo Plateau and Bogong High Plains, Victoria; Consuelo Plateau, Dicks Tableland and parts of the Atherton Tableland, Queensland; Ben Lomond Plateau, Tasmania.

Heights in feet have been given for the mountains in the following tables (3.2 – 3.6). Altitudes in feet are still used by the aviation industry, and many people feel that a measurement in metres for mountains lacks a certain aesthetic quality. The imperial measurement of the foot has historical roots relating to the physical world, unlike the metric measurement, which is purely scientific.

Hamersley Range escarpment, WA

Hamersley Range tableland, WA

Table 3.1 Highest Mountain in each State/Territory

NSW	Mt Kosciuszko, Snowy Mountains	2228m
Vic	Mt Bogong, Victorian Alps	1986m
Qld	Mt Bartle Frere, Bellenden Ker Range	1611m
SA	Mt Woodroffe, Musgrave Ranges	1439m
WA	Mt Meharry, Hamersley Ranges	1251m
Tas	Mt Ossa, Central Plateau	1617m
NT	Mt Zeil, Macdonnell Ranges	1531m
ACT	Bimberi Peak, Brindabella Ranges	1913m
OSA	Mawson Peak (Big Ben), Heard Island	2745m

Cross-sections of each of the highest mountains can be seen in figure 3.1. The highest mountain of the Australian Capital Territory, Bimberi Peak, shares a common border with New South Wales. The highest mountain wholly within the Australian Capital Territory is Mt Boyle at 1791m.

Summit of Mt Kosciuszko, NSW

Table 3.2 The '7000' Footers
Peaks and Tops over 7000 Feet in New South Wales

Mt Kosciuszko	7308ft	2228m
Mt Townsend	7246ft	2209m
Mt Twynam	7200ft	2195m
Rams Head	7183ft	2190m
Mt Etheridge	7183ft	2190m
unnamed peak on Etheridge Ridge	7150ft	2180m
Rams Head North	7141ft	2177m
Mt Lee	7098ft	2164m
Mt Alice-Rawson	7085ft	2160m
unnamed peak near Abbott Peak	7082ft	2159m
unnamed peak near North Rams Head	7045ft	2148m
Abbott Peak	7036ft	2145m
Carruthers Peak	7036ft	2145m

All the mountains and hill summits listed in table 3.2 lie within the Snowy Mountains region on the Kosciuszko Plateau. Many tops rise only a few hundred metres above the general level of the plateau upon which they stand. The western fall of this plateau forms the highest hillslope in Australia. Even though Australia's mountains are not really high, the effect of the relatively rarefied air can certainly be felt by anyone undertaking physical activity there; a shortness of breath is common until one adjusts.

Table 3.3 The '6000' Footers
Peaks and Tops over 6000 Feet in Victoria

Mt Bogong	6516ft	1986m
Bogong West Peak	6447ft	1965m
Mt Feathertop	6306ft	1922m
Mt Nelse North Peak	6178ft	1883m
Mt Fainter South Peak	6158ft	1877m
Mt Loch	6148ft	1874m
Mt Hotham	6101ft	1860m
Mt Niggerhead	6046ft	1843m
Mt McKay	6043ft	1842m
Mt Cope	6027ft	1837m
Mt Fainter North Peak	6027ft	1837m
Spion Kopje	6025ft	1836m
Mt Cobberas	6025ft	1836m

The summits in table 3.3 are all found within the Victorian Alps, mostly in the vicinity of the Bogong High Plains, which is actually a plateau. They barely rise 100m above the plateau's surface. Mt Feathertop on the other hand is a striking mountain, especially when snow-covered in winter. When viewed from the west near Bright, this is considered to be the most 'alpine-looking' mountain on the Australian mainland. Mt Bogong is also impressive when seen from the Kiewa Valley.

On the next page are the highest mountains of the other states – only major peaks, rather than multiple summits, have been listed.

Mt Feathertop from Mt Hotham, Vic

HILLS, MOUNTAINS AND TABLELANDS

Table 3.4 The Highest Mountains of the Other States

Queensland
1 Bartle Frere South Peak, Bellenden Ker Range	5320ft	5320ft
2 Bellenden Ker South Peak, Bellenden Ker Range	5220ft	1591m
3 Mt Carbine, Main Coast Range	4650ft	1417m

South Australia
1 Mt Woodroffe, Musgrave Ranges	4707ft	1435m
2 Mt Charles, Mann Ranges	4370ft	1332m
3 unnamed peak near Amata, Musgrave Ranges	4244ft	1294m

Western Australia
1 Mt Meharry, Hamersley Range	4110ft	1253m
2 Mt Bruce, Hamersley Range	4051ft	1235m
3 Mt Frederick, Hamersley Range	4051ft	1235m

Tasmania
1 Mt Ossa, Central Plateau	5305ft	1617m
2 Legges Tor, Ben Lomond Plateau	5160ft	1573m
3 Mt Pelion West, Central Plateau	5150ft	1570m

Northern Territory
1 Mt Zeil, Macdonnell Ranges	5022ft	1531m
2 Mt Edward, Belt Range	4668ft	1423m
3 Mt Giles, Macdonnell Ranges	4556ft	1389m

Australian Capital Territory
1 Bimberi Peak, Brindabella Ranges	6276ft	1913m
2 Mt Gingera, Brindabella Ranges	6084ft	1855m
3 Mt Murray, Brindabella Ranges	6053ft	1845m

Mt Zeil, NT

Table 3.5 Highest Mountains: Australian Islands

Mawson Peak (Big Ben), Heard Island	9002ft	2744m
Mt Bowen, Hinchinbrook Island, Qld	3750ft	1143m
Mt Gower, Lord Howe Island, NSW	2838ft	865m
Strzelecki Peak, Finders Island, Tas	2549ft	777m
Mt Lidgbird, Lord Howe Island, NSW	2503ft	763m
Mt Munro, Cape Barren Island, Tas	2349ft	716m
Mt Maria, Maria Island, Tas	2326ft	709m
Mt Bruny, Bruny Island, Tas	1883ft	574m
Balls Pyramid, Lord Howe Island Group, NSW	1840ft	561m
Mt Hamilton, Macquarie Island, Tas	1420ft	433m

So far as table 3.5 is concerned, many of Australia's islands are rugged and mountainous. Those near the mainland coastline can be considered to be formerly part of the mainland during times of lower sea levels, the intervening valleys being flooded by rising seas after the last ice age.

Mt Bruce, WA

Table 3.6 The Heights of some Well-Known Peaks

New South Wales

Peak	ft	m
Round Mountain, Northern Tablelands	5202ft	1586m
Point Lookout, Northern Tablelands	5130ft	1564m
Mt Barrington, Barrington Tops	5101ft	1555m
Mt Kaputar, Nandewar Range	5000ft	1524m
Mt Canoblas, Central Tablelands	4580ft	1396m
Mt Exmouth, Warrumbungle Range	3956ft	1206m
Mt Warning, North Coast	3750ft	1143m
Mt King-George, Blue Mountains	3471ft	1058m
Mt Imlay, South Coast	2904ft	885m
Mt Gibraltar, Southern Tablelands	2831ft	863m
Pidgeon-House, Budawang Range	2362ft	720m
Cambewarra Mountain, Cambewarra Range	2050ft	625m

Mt Ka

Victoria

Mt Buller, Great Dividing Range	5927ft	1807m
The Horn, Mt Buffalo Plateau	5653ft	1723m
Mt Wellington, Howitt High Plains	5269ft	1606m
Mt Baw Baw, Baw Baw Plateau	5127ft	1563m
Mt William, The Grampians	3829ft	1167m
Mt Macedon, Great Dividing Range	3317ft	1011m
Mt Buangor, Great Dividing Range	3247ft	990m
Mt Latrobe, Wilsons Promontory	2475ft	755m
Mt Dandenong, Dandenong Range	2077ft	633m

The Horn, Mt Buffalo Palteau

Queensland

Peak	ft	m
Thornton Peak, Thornton Range	4511ft	1375m
Mt Dalrymple, Clarke Range	4190ft	1277m
Mt Lindesay, McPherson Range	3914ft	1193m
Consuelo Peak, Carnarvon Range	3851ft	1174m
Mt Kiangarow, Bunya Mountains	3759m	1146m
Mt Cordeaux, Great Dividing Range	3724ft	1135m
Mt Beerwah, Glasshouse Mountains	1824ft	556m

Glasshouse Mountains

South Australia

Peak	ft	m
Mt Morris, Musgrave Ranges	4215ft	1285m
Mt Whinham, Mann Ranges	4027ft	1228m
St Marys Peak, Flinders Ranges	3822ft	1165m
Rawnsley Bluff, Flinders Ranges	3199ft	975m
Mt Remarkable, Flinders Ranges	3150ft	960m
Mt Bryan, Mid-North	3057ft	932m
Mt Lofty, Mt Lofty Ranges	2384ft	727m

Flinders Ranges

Mt Augustus, WA

HILLS, MOUNTAINS AND TABLELANDS

Western Australia

Mt Stevenson, Hamersley Range	3844ft	1172m
Mt Augustus, Gascoyne	3625ft	1105m
Bluff Knoll, Stirling Ranges	3596ft	1096m
Mt Ord, King Leopold Range	3070ft	936m
Peak Charles, Yilgarn	2160ft	658m
Twin Peaks, The Porongurups	2145ft	654m
Mt Frankland, The South-West	1384ft	422m

Tasmania

Cradle Mountain, Central Plateau	5065ft	1544m
Frenchmans Cap, The South-West	4756ft	1450m
Mt Anne, The South-West	4675ft	1425m
Mt Wellington, near Hobart	4165ft	1270m
Federation Peak, The South-West	4009ft	1222m
Precipitous Bluff, The South-West	4000ft	1219m
Millers Bluff, Central Plateau	3977ft	1212m

Northern Territory

Mt Sonder, Macdonnell Ranges	4526ft	1380m
Mt Razorback, Macdonnell Ranges	4179ft	1274m
Mt Hay, Macdonnell Ranges	4107ft	1252m
Mt Olga (Kata Tjuta), Central Australia	3507ft	1069m
Ayers Rock (Uluru), Central Australia	2845ft	867m
Mt Conner (Atila), Central Australia	2833ft	863m
Central Mt Stuart, Central Australia	2770ft	844m

Ayers Rock

Australian Capital Territory

Mt Kelly, Scabby Range	5999ft	1829m
Mt Gudgenby, Scabby Range	5703ft	1739m
Mt Franklin, Brindabella Ranges	5400ft	1646m
Tidbinbella Peak, Tidbinbilla Range	5300ft	1615m
Mt Ainslie, Southern Tablelands	2763ft	842m

Mt Ainslie

Table 3.6 has been compiled, mostly, on the basis that each of the peaks and tors generally rise well above their surrounding lowlands and are hence a prominent feature of the landscape. Mt Olga, Central Australia, for instance, rises 546m above the plain, Ayers Rock (Uluru) rises 348m and Mt Conner, 300m.

DID YOU KNOW?

Distinctive Mountains

Australia's **highest freestanding mountain** is probably Walshs Pyramid near Gordonvale, Queensland – it rises about 910m above the surrounding lowlands. A freestanding mountain is one unattached to other mountains or ridges. Prominent peaks that give the appearance of being free-standing from certain angles include: Mt Kaputar, New South Wales, which rises 1300m on its western side above the plain; Bluff Knoll in the Stirling Ranges, Western Australia, rising about 950m above the plains; the very grand Mt Augustus, Western Australia. Mt Augustus was previously considered to be the world's largest rock but it is not really a single rock like Ayers Rock (Uluru); rather it is a free-standing and partially vegetated fold or monocline in the earth's crust. Nevertheless, it is big, and stands 858 metres above the broad Lyons River valley – as such it is the world's largest monocline. Ayers Rock is **world's largest rock**.

Wilsons Peak, McPherson Range, New South Wales-Queensland, standing 1231m high, is unique in Australia in as much that rain falling upon its summit can potentially flow into three different States: via the Clarence River, New South Wales; via various creeks to Moreton Bay, Queensland; via Queensland's Condamine River to eventually reach the Murray River in South Australia. The people of Wycheproof, Victoria, lay claim to the **world's lowest mountain** – Mt Wycheproof, which is 149m high!

The Human Dimension - Occupying the High Country

Although the Australian high country is only of moderate elevation, it is still significant in such a flat land in as much as mountainous uplands and tablelands locally modify climate and weather patterns, providing cooler climates for particular agricultural, grazing, hydro-electrical and recreational pursuits. These uplands support numerous local environments and ecological niches, and are the major water catchment areas.

Tables 3.7-3.16 (following) show the highest located structures, buildings, settlements, towns, farms, roads and railways in Australia and in each State/Territory.

Mt Lofty

Table 3.7 Highest-Located Structure/Building in each State/Territory

NSW	Seamans Hut, near Mt Kosciuszko	2035m
Vic	Summit Hut, near Mt Bogong	1940m
Qld	Bellenden Ker Tower, near Cairns	1593m
SA	probably on Mt Lofty, Mt Lofty Ranges	727m
WA	Mt Nameless Tower, near Tom Price	1126m
Tas	huts on Legges Tor, Ben Lomond Plateau	1560m
NT	tower above Heavitree Gap, near Alice Springs	750m
ACT	Mt Ginini Tower, Brindabella Ranges	1762m

The human-made structure actually located at the greatest altitude in Australia is the cairn on top of Mt Kosciuszko, situated at 2220m. A toilet is being planned to be built near Mt Kosciusko, probably at Rawson Pass, situated at 2123m. The Summit Hut at Mt Bogong has been destroyed by fire.

In a building constructed on top of Mt Kosciuszko by the meteorologist Clement Wragge (known as 'Inclement' to his friends), he and other meteorologists recorded the weather and other atmospheric conditions between 1897 and 1901. In winter the temperatures fell to -15°C and winds up to 225kms/hr were recorded.

Seamans Hut, NSW

HILLS, MOUNTAINS AND TABLELANDS

Blue Cow, NSW

Cabramurra, NSW

Table 3.8 Australia's Highest Settlements

Thredbo-Crackenback, alpine service area, NSW	1957m
Blue Cow, alpine village, NSW	1880m
Charlottes Pass, alpine village, NSW	1758m
Mt Hotham, alpine village, Vic	1750m
Perisher Valley, alpine village, NSW	1730m
Smiggin Holes, alpine village, NSW	1675m
Wire Plain, alpine service area, Vic	1650m
Guthega, alpine village, NSW	1640m
Mt Selwyn, alpine service area, NSW	1590m
Dinner Plain, alpine village, Vic	1585m
Falls Creek, alpine village, Vic	1580m
Mt Buller, alpine village, Vic	1560m
Mt Baw Baw, alpine village, Vic	1524m
Diggers Creek, alpine hotel, NSW	1520m
Tatra Inn, alpine hotel, Vic	1510m
Cabramurra, small town, NSW	1460m
Ben Lomond, alpine village, Tas	1440m
Dingo Dell, alpine service area, Vic	1410m
Kiandra, old ghost township, NSW	1396m
Wilsons Valley, alpine hotel, NSW	1372m
Thredbo, small alpine town, NSW	1370m
Ben Lomond, farming township, NSW	1363m
Ebor, farming township, NSW	1348m
Lake Mountain, alpine service area, Vic	1340m
Mt Buffalo, alpine hotel, Vic	1331m
Black Mountain, farming township, NSW	1320m
Guyra, country town, NSW	1319m
Llangothlin, farming township, NSW	1284m
Mt St Gwinear, alpine service area, Vic	1280m
Ebor, farming township, NSW	1260m
Shooters Hill, farming locality, NSW	1250m
Mt Stirling, alpine service area, Vic	1250m
Mt Donna Buang, alpine service area, Vic	1250m
Smokers Gap, alpine service area, ACT	1240m
Black Springs, farming township, NSW	1220m
Sawpit Creek, national park settlement, NSW	1200m
Mt Field, alpine service area, TAS	1200m
Old Adaminaby, tourist settlement, NSW	1200m

Table 3.8 includes all settlements and service areas over 1200m in altitude. Alpine service area denotes limited and temporary services available to skiers; alpine village denotes seasonal (maybe annual) services and accommodation.

Though inaccessible by public road, Thredbo-Crackenback is accessible by chairlift while Blue Cow is accessible by a rack railway. During winter months Charlottes Pass is only accessible by oversnow transport (and skiing). Mt Hotham is the highest settlement accessible by road all year.

DID YOU KNOW?

Australia's highest town is Cabramurra. **Australia's highest country town** is normally considered to be Guyra even though it is beaten by the farming townships of Ben Lomond and Black Mountain. By way of comparison the **highest Aboriginal campsite** yet discovered lies on a saddle below Perisher Gap, in the Snowy Mountains, New South Wales, at an altitude of 1830m.

Mt Buller, Vic

Thredbo

3.9 Australia's Highest Towns and Townships

Cabramurra	1460m
Thredbo	1370m
Ben Lomond	1363m
Black Mountain	1320m
Guyra	1319m
Llangothlin	1284m
Ebor	1260m
Black Springs	1220m
Old Adaminaby	1200m
Glencoe	1158m
Oberon	1105m
Glen Innes	1073m
Nimmitabel	1067m
Bell	1067m
Blackheath	1065m

These top fifteen highest towns and townships of Australia are all located on the tablelands and mountains of New South Wales. Thredbo has been included here for it offers services beyond those offered by alpine villages.

Table 3.10 Highest Settlement of each State/Territory

NSW	Thredbo-Crackenback, alpine service area	1957m
Vic	Mt Hotham, alpine village	1750m
Qld	Evelyn Central, locality	1085m
SA	Mt Lofty, tourist complex	727m
WA	Tom Price, mining town	740m
Tas	Ben Lomond, alpine service area	1440m
NT	Areyonga, Aboriginal settlement	672m
ACT	Williamsdale, roadhouse	745m

Evelyn Central, QLD

Table 3.11 Highest Town or Township in each State/Territory

NSW	Cabramurra, small hydro-electricity township	1460m
VIC	Bendoc, small farming and timber township	840m
QLD	Ravenshoe, farming town	910m
SA	Blinman, old mining township	616m
WA	Tom Price, mining town	740m
TAS	Miena, small township	1037m
NT	Alice Springs, regional centre	579m
ACT	Torrens, Canberra suburb	700m

The figure for Torrens, in Canberra, is approximate; also, its altitude may be exceeded by newer suburbs.

Determining the highest town in Victoria was not easy, owing to the fact that the interesting hill station town of Mt Macedon ranges in altitude from 550m to 860m. Its upper portions are therefore higher than Bendoc, but the 'centre' of Mt Macedon is about 620m above sea level and so it is beaten by Stanley (760m), Lyonville (760m), Benambra (729m) and Omeo (680m), as well as Trentham, Tolmie and Bullarto. In East Gippsland the Seldom Seen store near Wulgulmerang is situated 900m above sea level, but this settlement hardly constitutes a township. Although offering little indication of a township today, the old mining settlement of Grant in Victorian Alps was 1140m above sea level.

Another old mining settlement in Victoria, Aberfeldy, was located at an altitude of 1090m.

In South Australia, Blinman is exceeded in altitude by the approximately 690m high Aboriginal community of Amata, in the Musgrave Ranges.

Blinman township, SA

HILLS, MOUNTAINS AND TABLELANDS

3.12 Highest Suburb of Capital Cities in each State/Territory		
NSW	Faulconbridge	447m
Vic	Olinda	580m
Qld	West Bardon	100m
SA	Crafers	580m
WA	Kalamunda	275m
Tas	Ferntree	420m
NT	Winnellie	30m
ACT	Torrens	700m

Deciding which suburbs to include in table 3.12 was arbitrary – just as long as it was near the capital city, within commuting distance, and not separated from the city centre by more than one 'green belt'.

Ben Lomond farmland, NSW

3.13 Highest Farm or Station in each State/Territory		
NSW	Beulah Homestead, near Ben Lomond	1433m
Vic	Treasure's Homestead, Dargo High Plains	1460m
Qld	Farms near Evelyn Central, Atherton Tablelands	1085m
SA	Farms on Mt Bryan, near Hallett	930m
WA	Juna Downs Station, Hamersley Ranges	760m
Tas	Wihareja Homestead, near Shannon	920m
NT	Bond Springs Station, Macdonnell Ranges	700m
ACT	Gudgenby Station, south of Tharwa	960m

Table 3.13 measures the altitude of homesteads, farms or station properties – figures are approximate.
  In summer, seasonal grazing occurs at altitudes higher than those listed above in Victoria, Tasmania and New South Wales (along the Gungarlin River west of Eucumbene). The highest abandoned homestead, at 1740m, was Katingal, near Mt Jagungal, in the Snowy Mountains. In the Australian Capital Territory the abandoned Boboyan Station, on Naas Creek, was located 1150m above sea level.

Table 3.14 Highest Roads in each State/Territory

New South Wales

Road	Altitude
Kosciuszko Summit Road, at Charlottes Pass	1830m
Smiggin Holes-Guthega Link Road, north of Smiggin Holes*	1710m
Kings Cross Loop Road, at Kings Cross	1610m
Khancoban-Kiandra Road, near Round Mountain	1600m
Alpine Way, at Dead Horse Gap	1590m
Point Lookout Road, at summit	1564m

Victoria

Road	Altitude
Alpine Road, near Mt Hotham*	1820m
Pretty Valley Pondage Road, near the Ruined Castle	1780m
Cope Road, near Mt Cope	1720m
Mt Buller Summit Road, at the summit	1676m
Howitt High Plains Road, near Snowy Range landing ground	1630m
Dargo High Plains Road, on the Dargo High Plains	1628m

Queensland

Road	Altitude
Tumoulin Road, near Kennedy Highway junction *	1162m
Kennedy Highway, south of Tumoulin turnoff	1150m
Farm road, near Evelyn	1150m
Herberton Road, near Kennedy Highway junction	986m
Amiens Road, near Pozieres	961m
East Evelyn Road, at Kennedy Highway junction	960m

South Australia

Road	Altitude
Stokes Hill Lookout Road, at summit	750m
Mt Lofty Summit Road, at summit	726m
Alligator Gorge Road, south-west of Wilmington	700m
Mt Bryan East Road, near Mt Bryan*	700m
Farm road, east of Hallett	700m
Summit Road, near Mt Lofty	685m

Altitudes listed in table 3.14 refer to individually named roads open to the public. Figures rounded to 'zero' are approximate. In South Australia the Mt Lofty Summit Road is a different road to the Summit Road near Mt Lofty.

In New South Wales, a national park management trail continues on from Charlottes Pass to Rawson Pass, reaching an altitude of 2123m. This route was formerly part of the Kosciuszko Summit Road, which originally ran to the summit, reaching an altitude of 2220m. Queensland's highest 'private' road, at 1292m, leads to a tower on the summit of Mt Wallum, west of Atherton. In South

HILLS, MOUNTAINS AND TABLELANDS

Western Australia
Karijini Drive, west of Marandoo*	860m
Bunjima Drive, west of Mt Vigor	850m
Tom Price-Paraburdoo Road, east of Tom Price	780m
Mt Bruce Access Road, near car park	780m
Mt Robinson Rest Area Access Road, at car park	780m
Great Northern Highway, near Karijini Drive and Mt Robinson	760m

Northern Territory
Hamilton Downs Access Road, south-west of New Well	850m
Arltunga Tourist Drive, near Ankala Hill*	820m
Mereenie Loop Road (Larapinta Drive), near Mereenie Gas Field	800m
Pinnacles Road, near Pinnacles Bore	750m
Cattlewater Pass Road, near Cattlewater Pass	750m
Lookout Road, just east of Arltunga	750m

Tasmania
Ben Lomond Summit Road, near Legges Tor	1460m
Mt Barrow Access Road, at the summit	1340m
Lake Highway, north of Great Lake*	1210m
Poatina Highway, near Starvegut Hill	1190m
Lake Augusta Road, near Lake Augusta	1150m
Marlborough Highway, at Lake Highway junction	1065m

Australian Capital Territory
Mt Ginini Summit Road, at the summit	1762m
Mt Franklin Road, north of Mt Ginini	1600m
Boboyan Road, south of Gudgenby*	1400m
Corin Dam Access Road, at Smokers Gap	1240m
Brindabella Road, at Piccadilly Circus	1200m

* Indicates the highest through road for that State/Territory

Australia are a number of higher private roads crossing Aboriginal land in the Musgrave-Mann ranges, including the Wintawata-Ulkiya Road (maximum height 770m south of Wintawata), the Amata-Pipalyatjara Road (750m east and west of Aparatjara, and west of Amata) and the Illintjitja-Angatja Road (750m). In Western Australia the Mt Nameless 4WD track, near Tom Price, reaches 1128m at the summit. In the Northern Territory the private Kintore Road near Liebig Bore reaches 765m in altitude.

Karijini Drive, west of Marandoo, WA

DID YOU KNOW?

In table 3.15 some altitudes (rounded to 'zero') are approximate. Heights refer to roads specifically named a highway. **Australia's lowest surface road**, at -10m below sea level, is on the Oodnadatta Track, near Curdimurka, South Australia. Sub-sea level roads are located in tunnels beneath Sydney Harbour and Cooks River, in New South Wales, and beneath the Yarra River in Melbourne, Victoria.

Table 3.15 Highest Highways in each State/Territory

NSW	Snowy Mountains Highway, near Bullock Hill	1500m
Vic	Omeo Highway, near Mt Willis	1350m
Qld	Kennedy Highway, near Tumoulin Road turnoff	1162m
SA	Barrier Highway, south of Hallett	650m
WA	Great Northern Highway, between Karijini Drive and Mt Robinson	760m
Tas	Lake Highway, north of Great Lake	1210m
NT	Stuart Highway, near 16 Mile Bore	729m
ACT	Monaro Highway, near Williamsdale	790m

DID YOU KNOW?

On the old Ghan Line west of Curdimurka, near Lake Eyre South, the line crossed a creek at an altitude of approximately -10m below sea level. At -17m, the **lowest railway in Australia** is in a tunnel under Cooks River, New Southern Railway Line, Sydney.

Table 3.16 Highest Railway in each State/Territory

NSW	Blue Cow Mountain Terminal, Skitube Line	1875m
Vic	near Bullarto, Daylesford Line	750m
Qld	north of Ravenshoe, Ravenshoe Line	972m
SA	near Belalie North, Port Pirie-Peterborough Line	632m
WA	near Juna Downs, Rosella-Yandicoogina Line	770m
Tas	near Guildford, Emu Bay Line	703m
NT	north of Alice Springs, Darwin Line	730m
ACT	west of Queanbeyan, Canberra Line	583m

The Daylsford and Ravenshoe Lines are, in part, still operational and tourist trains operate over sections of these tracks. All other railway lines listed in table 3.16 were operational at the time of writing. The altitudes for Victoria, Western Australia and the Northern Territory are approximate.

The highest New South Wales Government railway line used to be on the now closed Main North Line south of Ben Lomond, Northern Tablelands, at 1376m above sea level – current highest is the Main North Line, north of Walcha Road, at 1102m above sea level. On the disused Newnes Line, in the Blue Mountains, a height of 1207m was reached. In Victoria the highest stretch of railway line was on the disused Cudgewa Line near Shelley (787m). In Tasmania a disused mining railway south of Rosebery attained an altitude of nearly 800m. Where the Canberra Line forms a common border with the Australian Capital Territory and New South Wales, it obtains a maximum altitude of approximately 800m near the Brooks Bank Tunnel.

Table 3.17 Highest Railway Station in each State/Territory

NSW	Blue Cow, Skitube Line	1875m
Vic	Bullarto, Daylesford Line	747m
Qld	Tumoulin, Ravenshoe Line	965m
SA	Belalie North, Port Pirie-Peterborough Line	632m
WA	Wallaroo, Kalgoorlie Line	500m
Tas	Guilford, Emu Bay Line	616m
NT	Alice Springs, Darwin Line	579m
ACT	Canberra, Canberra Line	561m

Table 3.17 refers to stations (Belalie North, Wallaroo and Guilford are more strictly sidings) in operation at the time of writing. Bullarto is a terminus on the Central Highlands Tourist Railway. Tumoulin is a terminus station as part of a tourist railway operating on the disused Herberton-Ravenshoe Line.

The highest government railway station in New South Wales was at Ben Lomond, Northern Tablelands, on the disused Armidale-Wallangarra section of the Main North Line, at an altitude of 1363m. In Victoria, the former highest railway station was at Shelley (781m) on the disused Cudgewa Line. In Western Australia the highest railway siding was at Pardoo (590m) on the disused Wiluna Line. In Tasmania the highest government railway station was at Stonor (445m). Royalla railway station, situated on the common border between the Australian Capital Territory and New South Wales, on the disused Queanbeyan-Bombala Line, was 792m in altitude.

Shelley railway station, Vic

HILLS, MOUNTAINS AND TABLELANDS

Volcanoes

Volcanic activity can raise the level of the land, either by producing distinctive cones or, which is more likely, by issuing lava out of vents. Lava flows bury pre-existing landscapes, forming volcanic plains. Extensive lava flows are located on the Keilor Plains west of Melbourne, across the Western Districts of Victoria, and throughout the North-East Highlands of Queensland, between Hughenden and Mt Surprise.

If volcanic lavas are particularly viscous, the volcanoes tend to build up into a characteristic cone. Volcanic cones can be seen at and around Mt Gambier, South Australia, Mt Eccles, Victoria, and on the Atherton Tableland, Queensland. After the volcano has become extinct and the cone has been subjected to a long period of erosion, all that might remain of it is the erosion-resistant volcanic plug standing high above the surrounding country. Such plugs are often spectacular features and can be seen in the Warrumbungles, New South Wales, as well as at mounts Warning and Kaputar, also in New South Wales, the Peak Range in central Queensland, or the Glasshouse Mountains of south-east Queensland.

Volcanoes are thought to be extinct in Australia today. The last volcanic activity on the Australian landmass occurred in the Mt Gambier area (Mt Gambier's last eruption was around AD600) in South Australia. The springs at Paralana in the northern Flinders Ranges, South Australia, where hot water bubbles up to the surface via a fault, give some indication of volcanic activity. Of course, on Heard Island in the Southern Ocean there is an active volcano - Big Ben. Nearby, the McDonald Islands also exhibit volcanic activity.

Table 3.18 Australia's Volcanoes

Active – Last Eruption
Big Ben, Heard Island	AD1992
McDonald Islands	AD 2001

Possibly Dormant
Mt Gambier, SA	AD540+90 years

Recently Extinct
Mt Schank, SA	1400 to 20,000 years ago
Mt Napier, Vic	1400 to 20,000 years ago
Mt Eccles, Vic	1400 to 20,000 years ago
Tower Hill, Vic	1400 to 20,000 years ago
Mt Burr, SA	20,000 years ago
McBride Volcanic Province, Qld	100,000 years ago

Ancient
Mt Canoblas, NSW	11M years ago
Barrington Tops area, NSW	12M years ago
Warrumbungles, NSW	16M years ago
Lord Howe Island/Balls Pyramid, NSW	17M years ago
Mt Kaputar, NSW	18M years ago
Mt Warning, NSW	21M years ago
Toowoomba area, Qld	23M years ago
Glasshouse Mountains, Qld	25M years ago

DID YOU KNOW?

Though in no way a volcanic feature, Australia does have possibly the **oldest and largest burning mountain in the world**. Near Wingen, New South Wales, there is an underground seam of coal that issues smoke and sulphurous odours through vents in the earth. Giving the illusion of volcanic activity, the fire has been burning for at least 5000 years.

Strzelecki Ranges, Vic

Earthquakes

The folding and faulting of the earth's crust (discussed earlier in the section on 'mountain building') usually happens very slowly. But minor adjustments that occur to release the tensions created can result in the rapid movements we experience as earthquakes. Minor earthquakes, known as tremors, may pass unnoticed. Australia regularly experiences earthquakes – 23 earthquakes, with a magitudinal range between 2 and 4.4, were recorded in the first two months of 2004 while a +6 magnitude earthquake can be expected once every 5 years. Australia experienced 201 +2.5 magnitude earthquakes in 2001. Values of magnitude are measured on the Richter Scale, which range from 0 up to 10 (total devastation).

The area around Lake George, New South Wales, and to the north, is the centre for many small earthquakes. The township of Dalton is probably the most earthquake-prone settlement in Australia although nearby Sutton could make that claim – they recorded 90 earthquakes between December 2001 and April 2002. Along the Paralana Fault in the Gammon Ranges, South Australia, very small seismic tremors are experienced almost daily, making this one of Australia's most earthquake-prone zones. The sound of the so-called 'rumblings of Arkaroo' heard in this district are the result of these tremors. In the Burakin area, north-east of Perth, Western Australia, over 18,000 earthquakes (minor tremors) occurred in 2001-02, making this Australia's most seismically active area.

Table 3.19 Some Large Australian Earthquakes

Year	Location	Magnitude
1941	Meeberrie, north of Geraldton, WA	7.2
1988	Tennant Creek, NT (three major quakes)	7.0
1968	Meckering, WA	6.7
1979	Meckering, WA	6.2
1954	Adelaide, SA	5.6
1961	Robertson-Bowral, NSW	5.6
1989	Newcastle, NSW	5.6
1983	Hawker, SA	5.1
2004	Lake Mackay, WA-NT	5.1
1973	Picton, NSW	5.0

Meckering, WA

The earthquakes listed in table 3.19 had the potential to cause significant damage to property with resultant loss of life. The extent of these consequences depends on the nearness of inhabited areas and buildings to the centre of the earthquake and on the type of ground beneath them, and whether the buildings are designed to deal with earthquakes. Overall however, the risk of injury or death caused by earthquakes in Australia is considered to be small.

The Tennant Creek earthquake resulted in some people selling up and moving from the town. Interestingly, it used to be thought that this area was stable enough to become a site for hazardous wastes. The 1968 Meckering earthquake caused extensive damage to buildings while the 1979 Meckering earthquake was significant in that it created a fault 1.5m wide and 37km long. The township of Meckering can expect a +4 magnitude earthquake once every five years. The Newcastle earthquake was significant in the loss of 13 lives, and the amount of damage it caused.

Table 3.20 Earthquake Hazard Potential in Australia

Meckering-Cadoux districts, WA
Tennant Creek district, NT
west of Lake Mackay, Great Sandy Desert, WA
parts of the Mt Lofty-Flinders Ranges, SA
Hervey Bay district, Qld
Poeppels Corner district, Simpson Desert, Qld-SA-NT
Lake George-Dalton district, NSW
Newcastle district, NSW
Millicent district, SA

Table 3.20 lists the risk posed by earthquakes to life and property in a decreasing order of potential hazard.

Meckering-Cadoux, WA

CHAPTER FOUR
THE LAND: GORGES, VALLEYS, CAVES AND CRATERS

THE LAND: GORGES, VALLEYS, CAVES AND CRATERS

Gorges and Valleys

Streams eroding into uplifted land generally form gorges and valleys. A valley is usually a long depression with fairly regular downward slopes, at the base of which is the stream that has carved the valley from the surface rocks. If you look at a valley in cross-section, you can see its sides are distinguished by hillslopes. These slopes may be very steep (even cliffed) and narrow, thus forming a gorge; or they might be steep but not so narrow thus forming a canyon, the side tributaries of which are sometimes known as ravines. Small valleys with moderately steep slopes are generally known as gullies; particularly attractive ones are sometimes called glens. Australia has many gorges, canyons and gullies on the edges of its tablelands.

Valleys mostly tend to be more open than those already described and are common across Australia, but not in plains country. Gentle hillslopes form broad valleys, the stream itself occupying a floodplain in the valley bottom. Eventually the slopes themselves may become so gentle that they form peneplains (see chapter 2).

Moderately steep slopes produce hilly or mountainous country. It is an arbitrary measurement – 1000ft (or approximately 310m) of local relief – that usually distinguishes a hill from a mountain in definitions. The term 'local relief' refers to the difference in height between a valley's bottom and an adjacent hillslope's summit. Mountains are important to some people. Not so long ago the people of Townsville wanted to artificially increase the height of Castle Hill in the centre of the city with quarried material, from 938ft (286m), just so that it topped the magic threshold of 1000ft to become Castle Mountain.

Not all valleys are formed by streams. In areas of faulting (see chapter 3) those parts of the earth's crust that have sunk below the general level of the land form a rift valley. Spencer and St Vincents gulfs in South Australia occupy rift valleys. In the same type of country where a lowland lies between two uplifted ranges, the lowland (itself perhaps uplifted but not as high) is called an intermontane valley or basin. There are such basins in the Flinders Ranges, South Australia. A type of valley sometimes found in limestone country is a dry valley, where the stream has disappeared, normally flowing underground through cave systems.

Cliff face, Kings Canyon, NT

Little River Gorge, Vic

Table 4.1 Deepest Gorge of each State/Territory		
NSW	Kanangra Deep, Blue Mountains	up to 830m
Vic	Little River Gorge, Victorian Alps	500m
Qld	Barron Gorge, Wet Tropics	400m
SA	probably along Alerumba Creek, Flinders Ranges	up to 300m
WA	probably unnamed gorge below Mt Frederick, Hamersley Ranges	330m
Tas	Fury Gorge, west of Cradle Mountain	760m
NT	Olga Gorge, Kata Tjuta, Central Australia	500m
ACT	Molonglo Gorge, Southern Tablelands	100m

When is a gorge not a gorge but rather a deep valley? Gorges are usually deep and narrow, which is the reason the canyons of the upper Blue Mountains, New South Wales, have been excluded from table 4.1. Measuring the depth of gorges is fraught with difficulty: while a gorge's bottom is readily determined, its upper levels are often uneven. The figures in tables 4.1 – 4.3 are approximate.

GORGES, VALLEYS, CAVES AND CRATERS

Table 4.2 Depths of Some Well-Known Gorges

Apsley Gorge, Northern Tablelands, NSW	up to 755m
Shoalhaven Gorge, Southern Tablelands, NSW	575m
Hillgrove Gorge, Northern Tablelands, NSW	490m
The Gorge (Mt Buffalo), Victorian Alps, Vic	480m
Wollomombi Gorge, Northern Tablelands, NSW	455m
Bungonia Gorge, Southern Tablelands, NSW	380m
Massey Gorge, Clarke Range, Qld	310m
Ediowie Gorge, Flinders Ranges, SA	up to 300m
Kings Canyon, Central Australia, NT	200m
Carnarvon Gorge, Central Highlands, Qld	200m
Murchison River Gorge, near Kalbarri, WA	180m
Standley Chasm, Macdonnell Ranges, NT	150m
Windjana Gorge, The Kimberley, WA	100m
Katherine Gorge, Top End, NT	60m

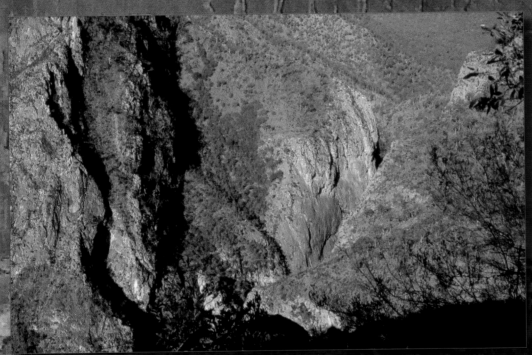

Bungonia Gorge, NSW

Table 4.3 Some Large and Deep Valleys

Geehi River Valley, Snowy Mountains, NSW	1330m
Lady Northcotes Canyon, Snowy Mountains, NSW	1000m
below Frenchmans Cap, The South-West, Tas	850m
Snowy River Valley, East Gippsland, Vic	820m
Herbert River Gorge, North-East Highlands, Qld	610m
Grose River Valley, Blue Mountains, NSW	600m
Cotter River Valley, Brindabella Ranges, ACT	400m

Some of the valleys and gorges mentioned in tables 4.2 and 4.3 can be seen in cross-section in figure 4.1.

Hillslopes and Cliffs

Though Australia has no really high mountains, it does have some impressive cliffs and hillslopes. Being a country of plains and plateaus it is on the eroded margins of our moderately high plateaus – or tablelands – that one can find our greatest hillslopes and highest cliffs.

Table 4.4 Greatest Hillslope in each State/Territory

NSW	western side of Mt Townsend, Snowy Mountains	1759m
Vic	north-western side of Mt Bogong, Victorian Alps	1651m
Qld	southern side of Mt Bartle Frere, Wet Tropics	1461m
SA	western side of Mt Brown, Flinders Ranges	961m
WA	south-eastern side of Bluff Knoll, Stirling Range	956m
Tas	western side of Precipitous Bluff, south-west Tas	1219m
NT	north-eastern side of Mt Zeil, Central Australia	831m
ACT	eastern side of Mt Gingera, Brindabella Range	848m

The comparative size of these hillslopes can be seen in cross-section in figure 4.2. Figure 4.3 shows cross-sections of hillslopes and valleys located near the state capitals.

Mt Bogong, Vic

GORGES, VALLEYS, CAVES AND CRATERS

Table 4.5 Some High Inland Cliffs and Near Vertical Hillslopes

New South Wales
below Watsons Crags, Snowy Mountains	500m
near Wollomombi Falls, Northern Tablelands	450m
below Point Lookout, Northern Tablelands	365m
below Bluff Mountain, Warrumbungle Range	300m
Kings Tableland, Blue Mountains	295m
some cliffs in the Blue Mountains	280m

Victoria
Little River Gorge, near Wulgulmerang	500m
The Gorge, Mt Buffalo Plateau	300m

Queensland
eastern side of Main Range, southern Qld	300m
near Wallaman Falls, Seaview Range	300m
Glass House Mountains, north of Brisbane	130m

South Australia
The Gorge, Mt Lofty Ranges	240m
below St Marys Peak, Flinders Ranges	200m
some gorges in the Gammon Ranges, Flinders Ranges	150m

Western Australia
below Bluff Knoll, Stirling Ranges	300m
some gorges in the Hamersley Range	100m

Tasmania
below Frenchmans Cap, south-west Tas	400m
below Federation Peak, south-west Tas	350m

Northern Territory
within Olga Gorge, Kata Tjuta, Central Australia	400m
Jim Jim Creek, Arnhem Land	200m

Australian Capital Territory
below Booroomba Rocks, Tidbinbilla Range	140m
above Rendezvous Creek, Tidbinbilla Range	100m

The heights in table 4.5 are approximate and exclude sea cliffs (see chapter 6). Near vertical hillslopes are included in this table for most cliffs are not necessarily sheer.

DID YOU KNOW?

The granite gorge below The Chalet at Mt Buffalo, Victoria is considered to be the highest **sheer cliff on the Australian mainland**, a claim sometimes given to the cliff at Bluff Knoll, Stirling Ranges, Western Australia, though it could be argued that this cliff is not sheer.

The cliffs below Frenchmans Cap, Tasmania, are the **highest sheer cliffs in Australia**. Worthy of note are the steep hillslopes of the Great Western Tiers, which rise 1000m above the bed of the Mersey River in northern Tasmania.

Mt Buffalo, Vic

Table 4.6 The Longest Cliffs of each State

NSW	cliffs of the upper Blue Mountains	280km
Vic	coastal cliffs between Princetown and Peterborough	30km
Qld	probably at Carnarvon Gorge, Central Highlands	30km
SA	Bunda Cliffs, Nullarbor Plain	210km
WA	Baxter Cliffs, Nullarbor Plain	165km
Tas	cliffs along Tasman Peninsula	over 25km
NT	cliffs of the Yambarran Range, Victoria River District	over 100km
ACT	Steamers Beach to Governor Head, Jervis Bay Territory	9km

These cliffs include coastal cliffs.

Port Campbell cliff face, Vic

The Human Dimension - Crossing the Range and Bridging the Gap

Traversing Australia's spectacular hillslopes are a number of roads. The following routes represent some of the most spectacular sections of roadway in Australia. Along these routes will be found very steep hillsides, cliffs, rock overhangs or awesome views. This list is by no means complete.

Table 4.7 Australia's Spectacular Roadways

New South Wales
Bilpin-Mt Irvine Road (officially closed), Blue Mountains
Jenolan Caves Road either side of the caves, Central Tablelands
Lawrence Hargrave Drive (currently closed) between Coalcliff-Scarborough, Illawarra
Wombeyan Caves Road between Bullio and Wombeyan Caves, Southern Tablelands

Victoria
Alpine Road between Dargo turnoff and Mt Hotham
Aberfeldy-Walhalla Road north of Aberfeldy River
Great Ocean Road between Lorne and Apollo Bay

Queensland
Cook Highway between Palm Cove and Port Douglas turnoff
Gillies Highway between Little Mulgrave and Lake Barrine

South Australia
Paralana Hot Springs track around Welcome Pound

Western Australia
Track to Mt Nameless summit, west of Tom Price
Stirling Ranges Drive, Stirling Ranges

Tasmania
Ben Lomond summit road near Jacobs Ladder
Lyell Highway between Derwent Bridge and Queenstown

The European settlement of Australia saw our transport networks – road and rail – having to bridge the gap across gorges or to climb the steep hillslopes of many of our ranges and tablelands. Such engineering constructions have resulted in a number

GORGES, VALLEYS, CAVES AND CRATERS

DID YOU KNOW?

The combined lengths of the Bunda and Baxter cliffs, rimming the southern edge of the Nullarbor Plain for a total 375km in length, make this **one of the longest continuous clifflines in the world**. It is possible that the discontinuous cliffs bounding the sandstone outcrops of Arnhem Land, Northern Territory, run for many hundreds of kilometres.

Alpine road below Mt Hotham, Vic

Table 4.8 Greatest Road Ascent of each State/Territory

NSW	Alpine Way, Tom Groggin-Dead Horse Gap, Snowy Mountains	1030m in 18kms
Vic	Alpine Road, Harrietville-Mt Hotham, Victorian Alps	1220m in 29kms
Qld	Gillies Highway, Little Mulgrave to The Bump, Wet Tropics	739m in 6.5kms
SA	Mt Lofty via Greenhill and Summit Roads, Mt Lofty Ranges	630m in 9kms
WA	Wittenoom Road, Python Pool to Mt Herbert, Pilbara	250m in 12kms
Tas	North Esk River to Ben Lomond summit, north-east Tas	1060m in 18kms
NT	not applicable	
ACT	Corin Dam Access Road, Tharwa Rd junction-Smokers Gap	580m in 10kms

These ascents are measured as the greatest height increase by a single road up a hillslope; they may include short downhill sections.

The greatest ascent is not in fact listed here, as it is not totally accessible to the motoring public. Between Swampy Plain Bridge, on the Alpine Way, and the top of Schlink Pass, north of Guthega, New South Wales, a road climbs 1380 metres in about 35km. This road is open to the public as far as Geehi Dam; beyond is a Snowy Mountains Authority management road. A place of interest along this road is Olsens Lookout, which offers awesome views of the Geehi River valley (Australia's deepest) and the immense hillside of the Main Range beyond (one of Australia's greatest hillslopes).

Steep Roads

One of the steepest ascents is on the Jenolan Caves-Oberon Road, Central Tablelands, New South Wales, just west of the Caves. This route rises 550 metres in 3.3 kilometres on an average gradient of 1 in 6. Other steep gradient roads include parts of the Maryville-Cumberland Junction Road, Central Highlands, Victoria (with a gradient of 1 in 6); northern access road through the Bunya Mountains National Park, Queensland (also 1 in 6); Sheoak Road, Belair, Mt Lofty Ranges, South Australia (1 in 4.5 – probably Australia's steepest road); Seymour Street, Albany, Western Australia (1 in 7.5) and Freestone Point Road, Triabunna, Tasmania (1 in 7).

Non-Urban Road Tunnels

Though they are becoming increasingly common along metropolitan expressways, road tunnels are uncommon in non-metropolitan Australia. The old Gwydir Highway, west of Grafton at Newton Boyd, New South Wales, has a 60.3m long tunnel while the Wombeyan Caves Road, west of Mittagong, New South Wales, has a 21.3m long tunnel at Bullio. On the Yelgun-Chinderah Freeway, part of the Pacific Highway, New South Wales, the new Cudgen road tunnel is 134m long. In the Mt Lofty Ranges, South Australia, are the new 500m-long twin bored Heysen Tunnels on the rerouted Adelaide-Crafers Highway. The Jenolan Caves Road, New South Wales, actually passes through the 137m-long Grand Arch Cave.

Some roads have made use of railway tunnels. Before the completion of the Gulgong-Muswellbrook railway a road (nowadays the Golden Highway) once passed through the 841m-long Coxs Gap No 1 Tunnel, while the disused 110m-long tunnel on the old Newnes Line, in the Blue Mountains, carries some 4WD traffic. In Queensland the 206m-long Boolboonda Tunnel on the Mt Perry Line is now used for road traffic.

Zig-Zags

The western ascent of the Blue Mountains replaced the famous Zig-Zag, a line cut into the side of the mountain in the pattern of a 'Z', the train travelling backwards down the slope of the 'Z'. Tourist trains travel this route today, passing through a tunnel and crossing attractive stone viaducts. Other Australian zigzags were used to ascend steep hills on the eastern side of the Blue Mountains near Lapstone, and near Kalamunda, Darling Range, Western Australia.

Railway Cuttings

There have been claims made that the 34m deep Windmill Cutting, near Toodyay, Western Australia, is Australia's deepest railway cutting. It is beaten by a cutting on the now disused Greenvale Line, Queensland (37.8m), and another cutting at Mt Bennett, Hamersley Line, Western Australia (40m). At 40.2m deep, Australia's deepest railway cutting is located near Zig Zag on the western ascent of the Blue Mountains. The maximum depth of the railway cutting on the old Picton-Mittagong Loop Line, New South Wales, is 23.7m.

By way of comparison, the deepest road cutting is on the Tumblong deviation, Hume Highway, south of Gundagai in New South Wales – depth: 49m. The Woakwine Cutting, near Robe, South Australia is 28m deep. Farmer Murray McCourt, with help, spent 3 years digging this 1km-long cutting to drain a swamp.

Embankments

The construction of railways has involved the use of some high embankments: Goonyella Line, west of Sarina, Queensland – 40.2m; on the Hamersley Iron Line in the Hamersley Ranges, Western Australia – 38m; on the disused Newnes Line, New South Wales – up to 30m.

Table 4.9 The Greatest Railway Ascent of each State

NSW	eastern side of the Blue Mountains, Main West Line	990m in 55kms
Vic	Bacchus Marsh-Ingliston, Western Line	356m in 21kms
Qld	Murphys Creek-Harlaxton, Main Line	396m in 27kms
SA	Clapham-Stirling West, Murray Bridge Line	460m in 21kms
WA	Pinjarra-Dwellingup, Dwellingup Line	254m in 22.5kms
Tas	Burnie-Goodwood Siding, Emu Bay Line	600m in 51kms
NT	not applicable	
ACT	not applicable	

In New South Wales the disused Newnes Line climbed nearly 400m in 15km, passing through 2 tunnels, on a steady 1 in 25 grade. In Victoria the greatest ascent used to be on the Cudgewa Line between Bullioh and Shelley (now disused); it rose 565m in 30.6km. In Queensland the Cairns-Kuranda Railway climbs 328m in 19km, passing through 15 tunnels and over numerous bridges in a 9km section.

The western ascent of the Blue Mountains, New South Wales, between Lithgow and Clarence, rises 138m in 12km. While not particularly great in terms of height achieved, the line does pass through 10 tunnels with a total length of 2,852.8m spread over 4.5km.

Steep Railways

The Skitube rack railway in the Snowy Mountains, New South Wales rises 755 metres in 8.5 kilometres. In Tasmania, the recently reopened Abt Railway between Strahan and Queenstown has a 4.8km rack section of 1 in 16 grades. A rack railway utilises a toothed rail over which a locomotive's cogwheel runs, enabling trains to ascend steeper gradients than by the mere adhesion obtained from a smooth wheel running over a smooth rail.

Railway Tunnels

To traverse steep hillslopes and ranges, railway tunnels were often required. Leaving aside metropolitan underground railway tunnels, the longest non-urban railway tunnels are presented in table 4.10.

Spiral Loops

To climb steep hills Australian railways have utilised two spiral loops: at Bethungra, Main South Line near Cootamundra, New South Wales, and the Border (or Cougal) Loop near the Queensland border, North Coast Line, New South Wales. A spiral loop is a line that passes over or under itself.

Bridges

In order to cross valleys and gorges, as well as streams and other water bodies, both railways and roads utilise bridges in order to maintain steady gradients. As a result some of these bridges have high clearances – see table 4.11.

GORGES, VALLEYS, CAVES AND CRATERS

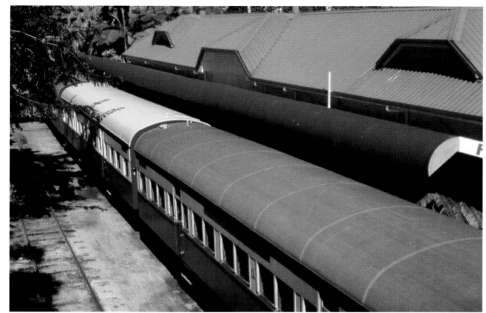

Kuranda railway station, Qld

DID YOU KNOW?

Though certainly not the greatest in terms of length, the Scenic Railway at Katoomba in the Blue Mountains is by far the steepest. This cable-hauled railway ascends 178 metres in 415 metres. The steepest gradient on the line is 52° from the horizontal, making it one of the **steepest inclined railways in the world**, though not the steepest as it is sometimes claimed.

Table 4.10 Longest Non-Metropolitan Railway Tunnels of Australia	
Bilson-Blue Cow Tunnel, Skitube, NSW	6300m
Bylong Tunnel, Gulgong-Muswellbrook Line, NSW	2032m
Woy Woy Tunnel, Main North Line, NSW	1789m
Border Tunnel, North Coast Line, NSW-Qld	1160m
Clifton Tunnel, Illawarra Line, NSW	1002m
Marrangaroo Tunnel, Main West Line, NSW	974m
Aylmerton Tunnel, Main South Line, NSW	919m
Rhydaston Tunnel, Hobart-Launceston Line, Tas	845m
Zig Zag No 10 Tunnel, Main West Line, NSW	789m
Monkerai Tunnel, North Coast Line, NSW	755m

Tunnel lengths are measured to the nearest metre. At one point the Bilson-Blue Cow Tunnel lies 550m below the ground surface. The once notorious Otford Tunnel, on the Illawarra Line, New South Wales, was 1550m long on a 1 in 40 grade – see figure 4.8. By way of comparison the longest tunnel in Victoria was on the privately owned Australian Portland Cement Company's narrow-gauge line at Geelong. It led from a quarry situated below sea level up a 1 in 37 grade through a 1,334m long tunnel. The longest non-metropolitan tunnel in Victoria today is the 422m long South Geelong Tunnel.

The longest non-metropolitan railway tunnels of the other states are: Kuranda Range No. 15 Tunnel, Cairns-Kuranda Railway, Queensland – 430m (longest urban tunnel: Brunswick Street Tunnel - 760m); Sleeps Hill Tunnel, Mt Lofty Ranges, South Australia – 728m; Pine Range No. 2 Tunnel, Goulburn-Canberra Line, Australian Capital Territory – 242.3m (shares common border with New South Wales). The only non-urban tunnel in Western Australia is the now disused Swan View Tunnel, Darling Range – 340m (longest urban tunnel: Subiaco Rail Tunnel – 900m). There are no railway tunnels in the Northern Territory.

The oldest tunnels in use are the Elphinstone (385m) and Big Hill (389m) tunnels on the Bendigo Line, Victoria, first used in 1862. The shortest tunnel (not a bridge) is the 40.2m long Croom Tunnel on the Illawarra Line, New South Wales.

Pheasant Nest Bridge, Wilton, NSW

Sydney Harbour Bridge, Sydney, NSW

Table 4.11 Highest Bridge Clearances of Australia

Bridge	Clearance
Pheasant Nest Bridge, Hume Freeway, Wilton, NSW	76m
Mooney Mooney Creek Bridge, Newcastle Freeway, near Peats Ridge, NSW	75m
Stanwell Creek Viaduct, Stanwell Park, NSW	65.5m (railway)
Nepean River Bridge, Hume Freeway, near Wilton, NSW	60m
Maldon Bridge, Wilton-Picton Road, near Maldon, NSW	60m (approx)
Gateway Bridge, Gateway Arterial Road, Brisbane, Qld	59m
Northbridge 'Suspension' Bridge, Cammeray, NSW	57m
Maribyrnong River Bridge, Keilor, Vic	56m
West Gate Bridge, West Gate Freeway, Melbourne, Vic	54.6m (railway)
Sydney Harbour Bridge, Bradfield Highway, Sydney, NSW	54m
Tasman Bridge, Tasman Highway, Hobart, Tas	53m
Spring Creek Bridge, near Tom Price, WA	46.2m
Moolgum Bridge, Hume Freeway, near Wilton, NSW	46m (railway)
Gladesville Bridge, Victoria Road, Drummoyne, NSW	43m
Werribee River Viaduct, Melton, Vic	41m
Back Creek Viaduct, Taradale, Vic	39m (railway)
Knapsack Gully Viaduct, lower Blue Mountains, NSW	37m (railway)
Woronora Bridge, near Sutherland, NSW	36.6m
	36m

In a few instances these figures represent bridge decking heights. For bridges passing over tidal waters, heights are measured against mean sea level.

Table 4.12 Some of Australia's Highest Bridge Structures

Bridge	Height
Sydney Harbour Bridge, NSW – top of arch	136.4m
Anzac Bridge, Sydney, NSW – top of tower	119.5m
West Gate Bridge, Melbourne – top of tower	102m
Batman Bridge, Tamar River, Tas – top of tower	96m
Story Bridge, Brisbane, Qld – top of structure	79.9m
Gateway Bridge, Brisbane, Qld – top of superstructure	64.5m
Gladesville Bridge, Sydney, NSW – top of structure	45.7m

DID YOU KNOW?

Australia's highest bridge, at 1,950m above sea level, is the Merritts Creek Bridge on the now-closed Kosciuszko Summit Road, New South Wales.

Caves

Caves are vacant spaces found on or near the surface of the land. Though not a significant feature of the landscape, a cave constitutes an important aspect of people's relationship to the land. Humankind's first home was probably the cave. Like the architecture of a building, a cave is a space that provided not only shelter from the elements but also, it could be argued, created an order to its occupier's life.

There are a number of different types of caves. There are rock shelters, which are common in rocky areas. They are formed by overhanging rocks in the base or face of a cliff, or by rocks fallen from cliffs or the jumbled tors of granite boulders. Tunnel caves on the other hand are created when streams erode weak joint chasms within a band of rock, in the process forming a cavity (a few rare examples can be found in the sandstone country surrounding Sydney, for instance, Walls Cave in the Blue Mountains, New South Wales). Lava caves or tunnels are another type, and are fairly common in areas that have experienced lava flows in the past. When the lava cools at different rates, such that some parts continue to flow while other parts have cooled and solidified, then lava caves such as the Undara Lava Tubes in northern Queensland are formed. Along rocky coastlines sea caves are found, having been formed by wave action eroding weak bands of rock. If a sea cave breaches the surface a short distance inland, blowholes may be formed. An unusual type of cave is located at Mt Gee in the northern Flinders Ranges, South Australia. Mt Gee is the only crystal quartz mountain in the world – its caverns and grottos are studded with crystal formations.

The most familiar caves are limestone caves, which can form within a whole bed of limestone, generally creating the most extensive cave systems and certainly the most complex. In simple terms, the formation of such caves starts when limestone is dissolved in water that contains carbon dioxide derived from the atmosphere and which is slightly acidic. The penetration of this acidic water enlarges minute joints or cracks within the limestone, which eventually causes the joints to enlarge. Sometimes the joints become so enlarged that the rock lying overhead may collapse into the cavity beneath. By these means the cave system grows and can cover many square kilometres and reach great depths. Limestone caves produce a multitude of forms: stalactites (growing down), stalagmites (growing up), sinkholes, collapsed dolines (that is, surfaces collapsing into the cave cavity beneath), underground streams and lakes and large caverns or arch caves. The best known of Australia's limestone caves are probably the Jenolan Caves in New South Wales, while one of the largest limestone belts in the world is found on and beneath the Nullarbor Plain.

Umpherstone Cave, Mt Gambier, SA

Grand Arch, Jenolan Caves, NSW

Table 4.13 Longest Cave in each State/Territory

NSW	Jenolan Caves system, Central Tablelands	approx 31km
Vic	B4 - Buchan Show Caves system, Gippsland	3km
Qld	The Queenslander, Chillagoe area	10km
SA	Corra-Lynn Cave, Yorke Peninsula	13.3km
WA	Old Homestead Cave, Nullarbor Plain	28km
Tas	Exit Cave, southern Tasmania	23km
NT	Bullita Cave system, Victoria River District	64km

The Bullita Cave system in Gregory National Park, Northern Territory, has multiple entrances making it maze-like rather than a single cave.

Table 4.14 Australia's Deepest Caves

Tasmania

Niggly Cave, Junee-Florentine area	375m
Anne-A-Kanada	373m
Growlling Swallet, Ice Tube system, Junee-Florentine area	360m
Khazad-Dum, Mt Field area	333m
Cauldron Pot, Junee-Florentine area	305m
Serendipty, Junee-Florentine area	278m

Mainland

Jenolan Cave system, Central Tablelands, NSW	180m
Eagles Nest Cave, Yarrongobilly system, Snowy Mountains, NSW	174m
Blowfly Cave, Bungonia system, Southern Tablelands, NSW	152m
The Bottomless Pit, Jenolan Cave system, Central Tablelands, NSW	149m

This list is not complete. Tasmania has many other caves deeper than those found on the mainland.

Table 4.15 Other Cave Facts

Combined total cave lengths:
Undara Lava Tubes, North-East Highlands, Qld   approx 160km

Biggest limestone room:
Abrakurrie Cave, Nullarbor Plain, WA   335m long, 45m wide, 30m high

Deepest single limestone natural shafts:
Niggly Cave, Junee-Florentine area, Tas   200m
Kellar Cellar, Mt Anne, south-west Tas   128m

Limestone caves generally form the most extensive cave systems and certainly the most complex, though lava tubes can be very long; the Undara Lava Tubes quoted above represent the total length of all passages. As cave exploration is still incomplete many of the above figures may well be exceeded during the currency of this book, especially in regard to depth and length.

GORGES, VALLEYS, CAVES AND CRATERS

Table 4.16 Some Australian Big Holes

The Crater, Mt Hypipamee National Park, North-East Highlands, Qld	140m
The Big Hole, Duea National Park, Southern Tablelands, NSW	96m
Devils Coachhouse, Jenolan Caves, Central Tablelands, NSW	85m
Kinrara Crater, North-East Highlands, Qld	80m
Great Nowranie Cave, Camooweal Caves system, Barkly Tableland, Qld	70m
Barrabarrac Hole, Gregory National Park, Victoria River District, NT	50m
Sparkes Gully, Port Campbell district, Western Districts, Vic	15m

This list is far from complete; in limestone country there are many big holes. The Crater at Mt Hypipamee is partially filled with water to a depth of 82 metres – making a clear 58m drop to water level. The Great Nowranie Cave 'hole' is 290m long. South of Wyndham, the Kimberley, Western Australia, is The Grotto, an unplumbed water-filled deep hole.

Table 4.17 Australian Meteorite Craters

Wolfe Creek, Tanami Desert, WA	1 crater: 853m dia. x 46m deep
Boxhole Crater, Central Australia, NT	1 crater: 175m dia. x 18m deep
Henbury Craters, Central Australia, NT	13 craters: largest - 157m dia. x 15m deep
Veevers Crater, Gibson Desert, WA	1 crater: 70m dia. x 8m deep
Dalgaranga Crater, Murchison Goldfields, WA	1 crater: 21m dia. x 3.2m deep

Sizes are approximate and this list is incomplete – many extraterrestrial-body impacts do not have the typical 'crater' form. Among these are included Lake Acraman, Gawler Ranges, SA (55km dia.); and Gosse Bluff, Central Australia, NT (22km dia./4km internal dia. x 250m deep).

DID YOU KNOW?

The Wolfe Creek Crater is considered to be the **world's 2nd largest meteorite crater**.

Undara lava tube, Qld

CHAPTER FIVE
THE WATER: STREAMS, WATERFALLS AND LAKES

THE WATER: STREAMS, WATERFALLS AND LAKES

Streams

Streams are the basic means of moving water over the surface of the land, and are found throughout much of Australia. The term stream includes what are commonly known as rivers and creeks, though these names can also apply to tidal estuaries. A stream normally contains the water flow in channels; flows operating above the general level of channels are known as floods.

Not all streams flow in channels, however. In some places, particularly in arid Australia, streams flow as sheets across the ground: parts of the upper reaches of the Gascoyne River in Western Australia do this. There are also braided streams, in which a multiplicity of channels intertwine with each other, and they are common in very flat country. The Channel Country of western Queensland is named after the braided channels of the Cooper, Diamantina and Georgina River systems. Another phenomenon associated with flat country streams is ana-branching: when their channels split and run off as smaller channels, perhaps to rejoin the main stream further on or perhaps even join another river system. An example of ana-branching can be seen in the lower reaches of the Murrumbidgee and Lachlan Rivers in the Riverina region of New South Wales. Split channels not rejoining the main stream (or another river system) are called distributaries. Willandra Creek, running off the Lachlan River in the Riverina, is a good example.

Stream in Bothwell, Tas

In Australia, many streams are perennial (flowing all year round), but most are ephemeral – they only flow after rain. Virtually all arid and semi-arid streams, as well as the tributary creeks in the humid and sub-humid regions, are ephemeral. In most cases however, even ephemeral streams retain water in all but the driest seasons in rockholes and waterholes. Waterholes are also called billabongs and are known as lagoons in northern Australia, pools in Western Australia and holes in Queensland.

Stream channels and their surrounding environs exhibit a number of interesting features. In upland or eroding areas can be seen waterfalls, cataracts, steep-sided gullies, rockholes, rapids, canyons, gorges and bluffs. Downstream, or in areas of deposition, there can be found floodplains, meanders, billabongs, terraces, deltas and alluvial fans.

STREAMS, WATERFALLS AND LAKES

5.1 Australia's Longest Rivers

King-Condamine-Balonne-Culgoa-Darling-Murray, Qld-NSW-SA	3749km
Murray, NSW	1870km
Barcoo-Cooper, Qld-SA	1600km
Murrumbidgee, NSW	1575km
Lachlan, NSW	1370km
Dawson-Fitzroy, Qld	1110km
Flinders, Qld	840km
Gascoyne, WA	760km
Burdekin, WA	710km
Murchison, WA	708km
Eateringinna-Alberga-Macumba, SA	630km
Goulburn, Vic	566km
Victoria, NT	560km
Fortescue, WA	547km
Roper, NT	418km

DID YOU KNOW?

Often considered (erroneously) as **Australia's longest river**, the Murray River's total length, from source to mouth, is 2520km. Sometimes Australia's longest river channel, based on maximum flows, is considered to be the Macintyre-Barwon-Darling-Murray with a length of 3370km.

Table 5.1 lists the longest singular stream channel within any one catchment; other long streams listed within that catchment are considered tributaries to the main stream. Hence, the Murray River is considered a tributary of the King-Condamine-Balonne-Culgoa-Darling-Murray stream channel. See figure 5.1 for the comparitive length of Australian rivers.

Obviously, singular channels have more than one name. It is possible that the Torrens-Cornish-Thomson-Cooper stream channel in Queensland and South Australia is longer than the Barcoo-Cooper stream channel listed above.

Table 5.2 Longest River in each State/Territory

NSW	Macintyre-Barwon-Darling-Murray	2771km
Vic	Goulburn	566km
Qld	Barcoo-Cooper	1110km
SA	Murray	650km
WA	Gascoyne	765km
Tas	South Esk	214km
NT	Victoria	560km
ACT	Naas-Gudgenby-Murrumbidgee	145km

These figures represent the total length of stream channels within each state. Though specific, these distances should be considered approximate - many were derived from old records. The Murray River has been excluded from Victoria, as the northern border of that state is the southern bank of the river.

Other long stream channels within each state include the Glenelg (457km) and Loddon (381km) in Victoria, the Archaringa-Neales (400km) in South Australia, the Gordon (185km) and Derwent (182km) in Tasmania, the Daly (362km) in the Northern Territory, and the Cotter (56km) and Murrumbidgee (51km) in the Australian Capital Territory.

South Australia, the driest state, has very few perennial streams: the longest one to rise within the state is the River Light, 161km long. Western Australia is mostly dry too: its longest perennial stream being the Blackwood River, 306km long. Tasmania has some longish rivers given the size of the state. If the Tamar Estuary is added to the South Esk River the total length increases to 271 kilometres.

Murray River, near Walkers Flat, SA

Table 5.3 Australia's Largest Catchment Areas

Murray-Darling, NSW-Vic-Qld-SA	1,062,530sq km
Cooper, Qld-SA	296,000sq km
Georgina-Eyre, Qld-SA-NT	242,000sq km
Diamantina-Warburton, Qld-SA	158,000sq km
Fitzroy, Qld	142,645sq km
Burdekin, Qld	129,860sq km
Finke, SA-NT	115,000sq km
Flinders, Qld	108,775sq km
Fitzroy, WA	88,980sq km
Murchison, WA	88,000sq km
Roper, NT	81,300sq km
Bulloo, NSW-QLD	78,000sq km
Victoria, NT	77,700sq km
Gascoyne, WA	77,600sq km
Ashburton, WA	76,700sq km

DID YOU KNOW?

Catchment areas, sometimes called drainage basins or just basins, are tracts of ground drained by a singular river system. Sometimes the catchment areas that drain into Lake Eyre, South Australia, are considered to be one drainage basin. Its combined area is 1,170,000 square kilometres, making it **Australia's largest internally drained catchment.**

The Human Dimension - Crossing the Bridge

The European occupation of Australia made use of the country's river systems and adjacent hinterland for grazing, agriculture and, prior to the advent of railways, transportation. Coastal sailing ships and steamers plied the seas and coastal rivers while paddle steamers coursed the larger streams of the Murray-Darling basin. At intervals settlements were established, often at fording points. With the establishment of road and railway networks, bridges were built to cross streams and other water bodies.

Burdekin Bridge, Qld

STREAMS, WATERFALLS AND LAKES

Sydney Harbour Bridge, NSW

Table 5.4 Australia's Longest Bridges	
Redcliffe Viaduct, Bramble Bay, Qld	2716m
Hornibrook Viaduct, Bramble Bay, Qld	2680m
West Gate Bridge, Yarra River, Vic	2582m
Western Distributor Viaduct-Anzac Bridge, Sydney, NSW	2088m
Sydney Harbour Bridge, NSW	2070m
Harris Park Viaduct, Sydney, NSW	1870m
West Gate Freeway Viaduct, Melbourne, Vic	1850m
Tasman Bridge, Derwent River, Tas	1417m
Gateway Bridge, Brisbane River, Qld	1627m
Story Bridge, Brisbane River, Qld	1375m
Gobba Bridge, Murrumbidgee River, NSW	1200m
Monash Freeway Viaduct, Melbourne, Vic	1197m
Sheehan Bridge, Murrumbidgee River, NSW	1143m
Burdekin Bridge, Burdekin River, Qld	1103m
Western Distributor Viaduct, Sydney, NSW	1045m
Macarthur Bridge, Nepean River, NSW	1030m
Stockton Bridge, Hunter River, NSW	1023m

The bridges listed in table 5.4 cross not only streams but also estuaries, embayments and, in the case of freeway constructions, built-up areas. All are road bridges. The figure for the Sydney Harbour Bridge is its total length including closed-in viaduct approaches - main bridge length is 1149m. The Gobba Bridge over the Murrumbidgee River is Australia's longest river crossing bridge (excluding river estuaries).

Causeways

Crossing the Thomson River floodplain at Longreach, Queensland, is a 6km long causeway incorporating 16 bridges and culverts - combined total bridge length is 2.2km. In Western Australia, between Point Peron and Garden Island, near Rockingham, is a combined bridge-causeway approximately 4.1km long. The Derwent River crossing at Bridgewater, Tasmania, is approximately 1100m long including about 700m of causeway; the Sorell Causeway-Bridge over Pitt Water, east of Hobart, is over 1460m long including a 460m long bridge. Other long bridge-causeways are located at the Mission River near Weipa, Queensland, between Karratha and Dampier, Western Australia, and near Port Hedland, Western Australia. Causeways are usually rock-filled embankments with occasional culverts or short bridges to allow for the movement of water.

West Gate Bridge, Vic

Table 5.5 Longest Road Bridge in each State/Territory

NSW	Sydney Harbour Bridge, Sydney	2070m
Vic	West Gate Bridge, Yarra River	2582m
Qld	Redcliffe Viaduct, Bramble Bay	2716m
SA	Swanport Bridge, Murray River	744m
WA	Mt Henry Bridge, Canning River	660m
Tas	Tasman Bridge, Derwent River	1417m
NT	Bradshaw Station access bridge, Victoria River	270m
ACT	Commonwealth Avenue Bridge, Lake Burley Griffin	310m

Other lengthy or well-known road bridges include:

New South Wales - Terragong Swamp Bridge, Minnamurra River (942m); the now closed Prince Alfred Bridge, Murrumbidgee River (921m); Hanwood Bridge, Clarence River (888m); Anzac Bridge, Blackwattle Bay (805m); Bethangra Bridge, Hume Reservoir (771m); Gladesville Bridge, Parramatta River (488m)
Victoria - Bolte Bridge, Yarra River (490m)
Queensland - Ross Camm Bridge, Pioneer River (620m)
South Australia - Berri Bridge, Murray River (330m); Hindmarsh Island Bridge, Goolwa (Lower Murray) River (319m)
Western Australia - Swan River Bridge, Swan River (397m); Fortescue Bridge, Fortescue River (397m); Dawesville Bridge, Dawesville Cut (360m)
Tasmania - Bowen Bridge, Derwent River (976m); Sorell Causeway Bridge, Pitt Water Estuary (460m); Batman Bridge, Tamar River (432m)
Northern Territory - South Alligator Bridge, South Alligator River (244m)

STREAMS, WATERFALLS AND LAKES

Table 5.6 Longest Railway Bridge in each State/Territory

NSW	Sydney Harbour Bridge, Sydney	2070m
Vic	Mitta Mitta River Bridge, Mitta Mitta River	878m
Qld	Burdekin Bridge, Burdekin River	1103m
SA	Algebuckina Bridge, Neales River	578m
WA	Bunbury Bridge, Swan River	413m
Tas	Bridgewater Bridge, Derwent River	310m
NT	Elizabeth River Bridge, Elizabeth River	n.a.
ACT	Jerrabomberra Creek Bridge, near Canberra	6.4m

Algebuckina Bridge, SA

DID YOU KNOW?

Australia's longest railway-only bridge, at 858m, is the Coomera River Bridge on the Brisbane-Gold Coast Line, Queensland.

The Mitta Mitta River and Algebuckina bridges are no longer in use. The figure for Tasmania is approximate. Both the Sydney Harbour and Burdekin River bridges are dual road/railway bridges.

The Wagga Wagga Viaducts over the Murrumbidgee River and floodplain have a total length of 3053m; it is composed of six viaducts (the longest approximately 1000m) joined by embankments. The Snowy River Trestle Bridges at Orbost, Victoria, have a total length of 769m including short embankments. The longest railway bridge in South Australia currently in use is the Murray River Bridge, which has a length of 573m. The Molonglo River Bridge, sitting astride the Australian Capital Territory-New South Wales border, is 108m long.

Waterfalls

Waterfalls are very scenic and Australia has, surprisingly for such a low and level country, some of the highest waterfalls in the world. This is because the rugged upland topography of Australia is created by streams eating away the edges of old tablelands and plateaus.

A waterfall is an interruption in the bed of a stream and it is commonly found on the edge of tableland country or where a band of hard rock cuts across the stream bed. There are a number of types of falls, the best known type being the true falls. This is where water falls over one or more sheer drops, at the base of which is a plunge pool. Lesser known are cataracts or cascades, in which the water flows over a succession of minor drops. But most common of all are rapids, where the stream flows over and down a bed or large and not-so large boulders and rocks. Due to the ephemeral nature of many of Australia's streams, some waterfalls are dry waterfalls and only flow after rain.

Determining the height of waterfalls is not easy. Things to consider are whether there is one of more drops, whether the drops are sheer, and whether to include upper level cataracts. The figures in table 5.7 include all drops including cataracts.

Fitzroy Falls, NSW

STREAMS, WATERFALLS AND LAKES

Table 5.7 Australia's Highest Waterfalls

Wollomombi Falls, Northern Tablelands, NSW	488m (multiple drops)
Wallaman Falls, Wet Tropics, Qld	305m
Dandongadale Falls, near Whitfield, Vic	255m
Tully Falls, Wet Tropics, Qld	300m (multiple drops)
Wulgulmerang Creek Falls, East Gippsland, Vic	245m (multiple drops)
Barron Falls, Wet Tropics, Qld	240m (multiple drops)
Jim Jim Falls, Kakadu, NT	200m
Apsley Falls, Northern Tablelands, NSW	196m (two drops)
Wentworth Falls, Blue Mountains, NSW	187m (multiple drops)
Fitzroy Falls, Southern Highlands, NSW	185m (two drops)
Bridal Veil Falls, Blue Mountains, NSW	180m
Katoomba Falls, Blue Mountains, NSW	165m
Ellensborough Falls, north coast of NSW	160m

Wallaman Falls is considered Australia's highest waterfall. The Ellensborough Falls is considered to be Australia's highest single drop waterfall. Other well known falls include: Steavensons Falls, Victoria - 82m; Serpentine Falls, Western Australia - 15m; Russell Falls, Tasmania - 40m.

Table 5.8 Highest Waterfall of each State/Territory

NSW	Wollomombi Falls, Northern Tablelands	220m (two drops)
Vic	Dandongadale Falls, near Whitfield	255m
Qld	Wallaman Falls, Wet Tropics	305m
SA	Morialta Falls, Mt Lofty Ranges	67m (two drops)
WA	King George Falls, The Kimberley	80m (two drops)
Tas	Montezuma Falls, near Rosebery	113m
NT	Jim Jim Falls, Kakadu	200m
ACT	Ginini Falls, Brindabella Ranges	approx 25m

The figures in table 5.8 generally refer to one or more drops, not necessarily sheer, but excluding upper level cataracts. Because waterfalls in South Australia rarely flow, it is hard to measure waterfall heights in that state; consequently, after rain, there may be higher falls in the northern Flinders Ranges. The same would be true of Western Australia and the Northern Territory, where there could be higher falls in the Kimberley and Hamersley Ranges, and at Ayers Rock (Uluru). It is possible the Revolver Creek Falls, Carr Boyd Ranges, the Kimberley, Western Australia, is higher than the King George Falls. In Western Australia the claimed greatest sheer drop is the 55m high Gordon Falls in the Hamersley Ranges; nearby, in the Ophthalmia Range, the Eagle Rock Falls is a 60m high cascade fall.

Morialta Falls, Mt Lofty Ranges, SA

Lakes

Australia is a land of many lakes, an unusual feature in such a dry continent, but less unusual when you consider that most of the lakes are dry. The Aborigines of south-west Queensland could tell whether or not a lake contained water by reading the sky. A soft lavender-blue rim of haze indicated a large expanse of open country, typically a lake bed. If water is present the rim is brighter, known as *kudje-oobra* or the telling of the water; if water is not present the rim is darker or called *wir-re-oobra*, the dust shadow. Australia has some of the largest in the world - saltlakes (for instance, Lake Eyre, South Australia) and claypans being the most common types. Most saltlakes are covered in a layer of salt while claypans are covered in a fine silt or clay; some dry lakes may have a thin covering of vegetation, however.

Other common lakes are found associated with river systems: these include billabongs (waterholes), ox-bows (billabongs occupying abandoned river meander loops) or flood-out or terminal lakes - for example, Bulloo Lake in north-western New South Wales (some saltlakes are also flood-out lakes). There are also overflow lakes - for instance, the Menindee Lakes in western New South Wales - and back-swamps, which are commonly found on riverine flood-plains. Vegetated back-swamps are perhaps better thought of as wetlands.

Some lakes result from volcanic activity: crater lakes, such as the Blue Lake at Mt Gambier, South Australia; caldera (exploded crater) lakes, such as Tower Hill in western Victoria; and maars (depressions intercepting the water table), such as Lake Corangamite, Victoria. Other lakes are formed by glacial action - when ice (or a glacier) deposits debris and rock materials (moraine) and scours the land surface. Cirque lakes such as Blue Lake near Mt Kosciuszko, New South Wales, are glacial lakes occupying ice-scoured depressions; while tarns - Hedley Tarn near Mt Kosciuszko, for instance - are formed by moraines damming creeks.

When the water table is intercepted, one can find bodies of water. These are common on the coastal plain around Perth during winter; on Fraser Island, Queensland, they are called window lakes. On Fraser Island, too, and the Great Sandy Region of Queensland there are perched lakes; these are located high on sand dunes and have a bed of impermeable, accumulated organic matter, which holds the water. Coastal dunes obstructing watercourses may form barrage lakes. Other coastal lakes found behind beaches or sand dunes are called lagoons.

Finally, there are the structural lakes, formed by various geological activities: for instance, Lake Karli Tarn, Victoria, created by a landslide damming a stream; or Lake George, New South Wales, which is a good example of a fault-angle lake created by adjacent geological strata. Most unusual are the underground lakes found in limestone country: a lake in Cocklebiddy Cave, Nullarbor Plain, Western Australia, is over 2.5km long.

The location of many of the lakes listed below can be seen in figure 5.1.

> **DID YOU KNOW?**
> **Australia's longest lake**, at 270km, is Lake Raeside, north of Kalgoorlie, Western Australia.

Table 5.9 Australia's Largest Natural Lakes

Lake	Area
Lake Eyre, SA	9500sq km
Lake Torrens, SA	5745sq km
Lake Gairdner, SA	4351sq km
Lake Mackay, WA-NT	3494sq km
Lake Frome, SA	2400sq km
Lake Macleod, WA	2380sq km
Lake Barlee, WA	1425sq km
Lake Moore, WA	1160sq km

The lakes in table 5.9 are all normally dry saltlakes. But Lake Eyre does receive water from time to time and has completely filled a few times over the past century.

Saltlake, near Skull Tanks, SA

STREAMS, WATERFALLS AND LAKES

The Inland Sea

Australia's fabled inland sea did actually exist within the past million years or so. Known as Lake Dieri, it covered the region occupied by present-day Lake Eyre and Lake Frome and points in-between, as well as the basins of the lower Cooper and Strzelecki creeks and Warburton River. With an approximate surface area of 100,000 sq km, it far exceeded today's largest lakes.

Table 5.10 Largest Natural Lakes of each State/Territory

New South Wales
Lake Cowal, floodout lake, near Burcher	162sq km
Lake George, fault-angle lake, near Collector	156sq km
Menindee Lake, overflow lake, near Menindee	155sq km

Victoria
Lake Corangamite, volcanic maar, near Coolac	209sq km
Lake Tyrell, floodout lake, near Sea Lake	172sq km
Lake Hindmarsh, floodout lake, near Jeparit	121sq km

Queensland
Lake Yamma Yamma, overflow lake, Channel Country	712 sq km
Bilpa Morea Claypan, claypan, north-east of Birdsville	475sq km
Lake Machattie, overflow lake, near Bedourie	310sq km

South Australia
Lake Eyre, saltlake, near Marree	9500sq km
Lake Torrens, saltlake, near Andamooka	5900sq km
Lake Gairdner, saltlake, near Kingoonya	4300sq km

Western Australia
Lake Macleod, saltlake, near Carnarvon	2380sq km
Lake Barlee, saltlake, near Youanmi	1425sq km
Lake Moore, saltlake, near Paynes Find	1160sq km

Tasmania
Great Lake, structural lake, near Miena	115sq km
Lake Sorrel, structural lake, near Interlaken	49sq km
Lake St.Clair, glacial lake, near Derwent Bridge	38sq km

Northern Territory
Lake Amadeus, saltlake, near Yulara	1032sq km
Lake Sylvester, floodout lake, Barkly Tableland	570sq km
Lake Woods, floodout lake, near Elliot	500sq km

Australian Capital Territory
Lake Windemere, coastal lagoon, Jervis Bay Territory approx	0.4sq km

Coastal lagoons are not taken into account in the compilation of table 5.10, except for Lake Windemere, Australian Capital Territory. In the case of claypans, floodout lakes and overflow lakes, sizes are approximate and may include adjacent floodplains. The sizes given in this table for Menindee Lake, New South Wales, and Great Lake, Tasmania, are their original size before alteration for hydrological works.

The figures in table 5.10 for New South Wales are exceeded by the outline of a huge lake with an area of 542sq km. Called Lake Garnpung, it is the bed of an ancient lake system that flowed from the prehistoric Lachlan River. The lake sizes given for Western Australia are exceeded by the large, border-straddling Lake Mackay (see table 5.9). In the Australian Capital Territory, the artificial Lake Burley Griffin is 7.2sq km.

Figure 5.2 shows the comparative sizes of the largest natural lakes of each State/Territory, except for the Australian Capital Territory, for which the lake is too small at this scale. For New South Wales, as both lakes Cowal and George are almost the same size, there is some question as to which is the larger: thus both are included.

DID YOU KNOW?

Great Lake today occupies 176sq km – it was, before alteration, considered to be **Australia's largest freshwater lake.**

Lake Eyre, Level Post Bay, SA

Table 5.11 Australia's Highest Natural Lakes

Lake Cootapatamba	2050m
Club Lake	1950m
Lake Albina	1915m
Blue Lake	1900m
Hedley Tarn	1860m

All the lakes in table 5.11 are of glacial origin and situated in the vicinity of Mt Kosciuszko, New South Wales, where there are also some smaller unnamed lakes or pools. In the same area, between Mt Townsend and Mt Alice Rawson, there used to be a glacial lake called Russel Tarn or Lake Claire. This lake was situated at an altitude of approximately 2150m but now seems to have completely dried up.

By way of comparison, the highest artificial lake is Pretty Valley Pondage, on the Bogong High Plains, Victoria, at 1630 metres above sea level.

Lake Cootapatamba, NSW

Table 5.12 The Highest Natural Lake of each State

NSW	Lake Cootapatamba, glacial tarn, Snowy Mountains	2050m
Vic	Lake Karli Tarn, structural lake, Victorian Alps	914m
Qld	Lake Eacham, crater lake, Atherton Tableland	755m
SA	Lake Wilson, claypan, Mann Ranges	680m
WA	Lake Kerrylyn, saltlake, Little Sandy Desert	560m
Tas	Menamatta Tarns, glacial tarns, Ben Lomond Plateau	1430m
NT	unnamed claypans, Missionary Plain	780m
ACT	not applicable	

The listing in table 5.12 is probably not complete. Apart from the dried-up Russel Tarn in the Snowy Mountains, New South Wales (see the comments to table 5.11 above), beneath the summit of Mt Ossa in Tasmania there is a small pool-sized tarn, at an approximate height of 1610m. The figure for the Northern Territory is approximate.

South Australia's highest permanent lake is Silver Lake, Mt Lofty Ranges, at a height of 290m. On the Central Plateau of Tasmania there are many glacial lakes situated above 1000m.

DID YOU KNOW?
Lake St Clair is **Australia's deepest lake.**

Table 5.13 Some Lake Depths

Lake St. Clair, Central Plateau, Tas	168m
Blue Lake, Lower South-East, SA	75m
Lake Eacham, Atherton Tableland, Qld	70m
Lake Bullenmerri, Western Districts, Vic	62m
Lake Barrine, Atherton Tableland, Qld	60m

The lakes in table 5.13 are probably representative of Australia's deepest lakes but the list is far from complete. Interestingly, the four shallower lakes are of volcanic origin while Lake St Clair, Australia's deepest lake, is of glacial origin. The Crater, at Mt Hypipamee, Queensland, is a partially filled hole with a water depth of 82 metres. Near Wyndham, Western Australia, is The Grotto, a deep rockhole: one measurement of its depth is over 120m, even though its surface area is about the size of a suburban house.

The Human Dimension - Some Hydrological Works

The continued European occupation of the Australian landmass has involved the establishment of hydrological works, to guarantee a continual water supply to service the needs of the population for drinking and other water uses — irrigation, agriculture, grazing, manufacturing, hydro-electricity generation and recreation. The construction of large dams and reservoirs can radically change a landscape. The impounded waters flood valleys, sometimes farmlands and settlements, destroy vegetation formations, change stream flow regimes and, in isolated country, impair or distract from the aesthetic values of wilderness.

Dams capture the runoff from water catchment areas (see table 5.3 above), storing water in reservoirs to be available for future use. Water is transported via a variety of means: natural stream channels, pipelines, aqueducts, tunnels and open channels. The longest single hydro-electrical scheme tunnel is the 23.4km-long Eucumbene-Snowy Tunnel in the Snowy Mountains of New South Wales. The longest linked hydro-electrical tunnel is the 49.7km-long Eucumbene-Snowy/Snowy-Geehi/Murray One Pressure Tunnel.

An unusual type of dam or, more correctly, weir, is the overshot. An overshot is a stone weir constructed across the downstream end of a waterhole that may have been partially excavated, thus creating an earth tank. The overshot (and excavation) raises the storage capacity of the waterhole (or tank), thus increasing the capture and storage of water after floods - this in turn provides stock with a more reliable water supply.

In many parts of rural Australia, groundwater is a significant source of supply. The Great Artesian Basin, a multi-layered sandstone aquifer, covers an area of 1.74 million square kilometres of New South Wales, Queensland and South Australia and has approximately 3100 active bores (down from a maximum of 4700). The bores are used for stock watering, mining and domestic purposes.

DID YOU KNOW?
At 2136m the Springleigh Bore, Great Artesian Basin, Queensland, is **Australia's deepest bore.**

Table 5.14 Australia's Largest Reservoirs

Lake Argyle, Ord River, WA	1076sq km
Lake Menindee, Darling River, NSW	458sq km
Lake Gordon, Gordon River, Tas	268sq km
Lake Pedder, Serpentine/Houn Rivers, Tas	242sq km
Lake Hume, Murray River, NSW-Vic	202sq km
Lake Pindari, Severn River, NSW	195sq km
Lake Dalrymple, Burdekin River, Qld	186sq km
Great Lake, Shannon River, Tas	176sq km
Lake Eucumbene, Eucumbene River, NSW	145sq km
Lake Maraboon, Nogoa River, Qld	144sq km
Lake Eildon, Goulburn River, Tas	138sq km
Lake Wivenhoe, Brisbane River, Qld	115sq km

There are nearly 500 major dams and reservoirs in Australia - a major dam is defined, in very general terms, as being more than 10m in height.

Table 5.15 Largest Reservoir in each State/Territory

NSW	Lake Menindee, Darling River	458sq km
Vic	Lake Eildon, Goulburn River	138sq km
Qld	Lake Dalrymple, Burdekin River	186sq km
SA	South Para, South Para River	4.4sq km
WA	Lake Argyle, Ord River	1076sq km
Tas	Lake Gordon, Gordon River	268sq km
NT	Darwin River, Darwin River	25.9sq km
ACT	Lake Burley-Griffin, Molonglo River	7.2sq km

Table 5.16 Australia's Highest Dams

Dartmouth Dam, Mitta Mitta River, Vic	180m
Thomson Dam, Thomson River, Vic	165m
Talbingo Dam, Tumut River, NSW	162m
Warragamba Dam, Wolindilly River, NSW	142m
Gordon Dam, Gordon River, Tas	140m
Reece Dam, Pieman River, Tas	122m
Eucumbene Dam, Eucumbene River, NSW	116m
Copeton Dam, Gwydir River, NSW	113m
Blowering Dam, Tumut River, NSW	112m
Cethana Dam, Forth River, Tas	110m

With the exception of the Warragamba and Gordon dams, all the dams listed in table 5.16 are earth and rock-filled dams. The Tumut Pond Dam, Tumut River, New South Wales is 86m high.

Eucumbene Dam, NSW

Table 5.17 Highest Dam in each State/Territory

NSW	Talbingo Dam, Tumut River	162m
Vic	Dartmouth Dam, Mitta Mitta River	180m
Qld	Split Yard Creek, Pryde Creek	76m
SA	Kangaroo Creek Dam, Torrens River	63m
WA	Ord Dam, Ord River	99m
Tas	Gordon Dam, Gordon River	140m
NT	Darwin River Dam, Darwin River	31m
ACT	Corin Dam, Cotter River	74m

Dartmouth Dam, Vic

Table 5.18 Australia's Longest Dams

Wurdee Buloc Dam, off stream, Vic	8670m
Ross River Dam, Ross River, Qld	8168m
Mokoan Dam, Winton Swamp, Vic	7500m
Kinchant Dam, Sandy Creek North Branch, Qld	5100m
Kununurra Diversion Dam, Ord River, WA	4663m
Claredon Dam, off stream, Qld	4300m
Pine Lake Dam, off stream, Vic	3700m
Beardmore Dam, Balonne River, Qld	2591m
Greenvale Dam, Yuroke Creek, Vic	2500m
Wivenhoe Dam, Brisbane River, Qld	2300m

In table 5.18, off stream refers to water storages not located on a stream.

Hot Springs

One does not normally associate Australia with hot springs. Hot spring waters emanate from within the earth's crust, where they have been heated. The origin of the heat may be volcanic (as, for example, at Tallaroo, Queensland, or Paralana, South Australia), or it may simply be that the spring's source lies deep enough within the earth's crust - for example, Dalhousie, South Australia.

The main Dalhousie hot spring is interesting inasmuch as it is just one of around 30 springs in the vicinity. The Dalhousie Springs are natural outlets of the Great Artesian Basin, as are Blanche Cup and The Bubbler, both in South Australia. These natural artesian springs are sometimes known as mound springs. This is because the sediment particles, which their waters carry to the surface, together with chemical precipitates, create mounds that are often quite a few metres high. Typically, the water flows from the top or sides of these mounds.

There are at least 90 mound springs, or complexes of springs, in South Australia alone. Many mounds are relict springs, having dried up in the past.

Not all Great Artesian Basin springs flow with water. Near Eulo in south-west Queensland, and elsewhere, are mud springs. These are mounds of dried mud; occasionally, when the underlying pressure becomes too great, they blow their tops.

Ross River Dam, Ross River, Qld (above and left)

Table 5.19 Some Australian Hot Springs

Yarrongobilly, Snowy Mountains, NSW	27°C
Innot, near Mt Garnet, Qld	80°C
Tallaroo, west of Mt Surprise, Qld	92°C
Blanche Cup, near Coward Springs, SA	25°C
Dalhousie, north of Oodnadatta, SA	43°C
Paralana, near Arkaroola, SA	95°C
The Bubbler, near Coward Springs, SA	25°C
Millstream, Pilbara, WA	25°C
Zebedee, near El Questro Station, WA	25°C
Thermal Pool, near Hastings, Tas	28°C
Douglas, Top End, NT	60°C
Mataranka, Top End, NT	31°C

The figures in table 5.19 are typical temperatures of spring waters either at point of issue or in nearby bathing facilities, and are approximate. The hot springs listed here are naturally occurring and should not be confused with artesian bores.

CHAPTER SIX
THE COAST

THE COAST

CHAPTER 6

The Coastline

Most Australians are familiar with coasts because most live near one - in fact about three out of four people live within an hour's drive of the sea. Around 9 out of 10 people live within 50 kilometres of the coast on a strip extending from Port Douglas in Queensland to Ceduna in South Australia, and around the south-west corner of Western Australia between Esperance and Geraldton.

While extensive bays and gulfs form the shape of the coastline, along the shores of Australia there are a number of different types of coasts to be found – such as beach coasts, rock coasts, tidal plain coasts or, off parts of the tropical coast, coral reefs and sandy cays. The most unusual Australian coast, however, occurs on Heard Island in the Southern Ocean. Here active glaciers have snouts, which override the land surface to form ice coasts.

THE COAST

DID YOU KNOW?

Australia's *most extensively bayed coast* is the Kimberley coast of Western Australia. Here numerous valleys have been flooded by rising sea levels since the last ice age; this has resulted in many bays, gulfs and numerous off-shore islands, a number of which form the Buccaneer and Bonaparte archipelagos.

Table 6.1 Coastline Length of each State/Territory

NSW	1900km
Vic	1800km
Qld	7400km
SA	3700km
WA	12500km
Tas	3200km
NT	6200km
ACT	35km
AUS	36735km

Gulfs and Bays

Embayments are indentations in the coastline. Bays are usually thought of as indentations of the land by bodies of water, in this case the sea or ocean. Gulfs are generally large embayments that penetrate far into the land. They are formed by rising sea levels flooding plains - for instance, the Gulf of Carpentaria; or by faults in the earth's crust resulting in a rift valley that is consequently flooded by the sea - for instance, Spencer Gulf and St Vincent Gulf.

TABLE 6.2 SOME AUSTRALIAN GULFS AND EMBAYMENTS: BY AREA

Gulf of Carpentaria, Qld/NT	335000sq km
Joseph Bonaparte Gulf, WA/NT	58000sq km
Spencers Gulf, SA	20000sq km
Shark Bay, WA	16500sq km
Van Diemen Gulf, NT	11500sq km
King Sound, WA	8200sq km
St Vincents Gulf, SA	5200sq km
Hervey Bay, Qld	4500sq km
Exmouth Gulf, WA	4200sq km
Moreton Bay, Qld	3300sq km
Admiralty Gulf, WA	2700sq km
Broad Sound, Qld	2700sq km
Port Phillip Bay, Vic	1950sq km
Blue Mud Bay, NT	1800sq km
Shoalwater Bay, Qld	1300sq km
Oyster Bay, Tas	600sq km
Roebuck Bay, WA	600sq km
York Sound, WA	500sq km
Macquarie Harbour, Tas	300sq km

There is some confusion over the length of the Australian coastline: the Year Book of Australia is quoted above. Other sources give different figures. A CSIRO study found that the length of the coastline was 30,270km including Tasmania and 47,070km if all islands are included, while the Australian Handbook quotes 19,320km and the Australian Encyclopaedia 19,650km. The different figures are due to different techniques in measuring coastlines – whether or not bays, estuaries or islands are included for instance.

This is not a complete list of the largest embayments: a few smaller ones have been included for comparison. The figures in table 6.2 are approximate, for there are no specific boundaries to embayments apart from coastlines. Although some of these places are named 'sound', they would perhaps be better termed bays or gulfs, since a sound is really a narrow water passage, like a strait.

Byron Bay, NSW

Gulf Carpenteria, Karumba, Qld

Table 6.3 Australian Capital City Embayments

St Vincents Gulf (Adelaide), SA	5200sq km
Moreton Bay (Brisbane), Qld	3300sq km
Port Phillip Bay (Melbourne), Vic	1950sq km
Derwent Estuary (Hobart), Tas	200sq km
Darwin Harbour (Darwin), NT	110sq km
Sydney Harbour (Sydney), NSW	55sq km
Swan-Canning Estuary (Perth), WA	53sq km
Botany Bay (Sydney), NSW	50sq km

Beach Coasts

Beach coasts have lengthy beaches, behind which there is usually a complex of dunes with perhaps minor features such as estuaries and lagoons - for instance, sections of the north coast of New South Wales. There is also a special type of beach coast, called a barrier beach coast, which has the lengthy beaches and dune complexes but is separated from the mainland by extensive lagoon systems – such as in the Gippsland Lakes district of Victoria or along The Coorong of South Australia.

Beaches are largely formed by wave action and are generally gently curved, making a concave shape with the shoreline. Beaches are generally composed of sand-sized particles with a range of colour variation that is dependent on the constituent minerals of the sands. The sands can range in colour from nearly white, common in the south-west of Western Australia (though the whitest beach sands are said to be at Hyams Beach, Jervis Bay, New South Wales) through to deep yellow, common on the east coast of Australia.

Not all beaches are sandy. Shingle beaches, composed of pebble-sized particles, are rare in Australia; however, Pebbly Beach, on the new South Wales south coast, is a good example. At Shark Bay, Western Australia, there is the aptly named Shell beach, composed of innumerable mollusc shells. In another type of beach, those fronting shores with high tidal ranges, broad foreshores of mud are revealed at low tide. Shallow temporary lagoons often form on these tidal flats as the tide ebbs. Such beach coasts can be seen at Western Port, Victoria, in Edgecombe Bay near Bowen, north Queensland and around Port Hedland, Western Australia.

Cactus Beach, SA

THE COAST

Ninety Mile Beach, Vic

Table 6.4 Longest Beach in each State/Territory

NSW	Newcastle Bight Beach, north of Stockton	30km
Vic	Ninety-Mile Beach, Gippsland	145km
Qld	Seventy-Five Mile Beach, Fraser Island	120km
SA	Younghusband Peninsula/Lacepede Beach, The Coorong	175km
WA	Eighty-Mile Beach, Great Sandy Desert	175km
Tas	Ocean Beach, near Strahan	34km
NT	Aurari Bay Beach, Arnhem Land	17km
ACT	Bherwerre Beach, *Jervis Bay Territory*	5km

Table 6.4 is probably not complete, for doubts exist about the longest beaches in New South Wales, Queensland, and the Northern Territory. As a matter of interest, Eighty-Mile Beach, Western Australia, should be called One Hundred and Ten-Mile Beach.

Distinctive of beach coasts are lagoons – coastal lakes partly or totally separated from the sea by narrow strips of land. This land could be sand spits, sand hills or, in the case of submerged coasts, low lying hills or ridges. Lagoons, if connected to the sea, will be tidal; their waters may be salty, brackish or fresh.

Table 6.5 Largest Lagoon of each State/Territory

NSW	Lake Macquarie, Central Coast	115sq km
Vic	Gippsland Lakes, Gippsland	380sq km
Qld	Noosa Lakes, Great Sandy Region	60sq km
SA	Lakes Alexandrina-Albert-Coorong, Lower South-East	970sq km
WA	Peel-Harvey Inlet, Coastal Plain	130sq km
Tas	Moulting Lagoon, near Swansea	50sq km
NT	unknown	
ACT	Lake Windemere, Jervis Bay Territory	0.4sq km

The sizes of these lagoons are approximate. Some of the lagoons named above are two or more lagoons joined by narrow waterways. These lagoons exclude coral lagoons or those inland waterholes called lagoons in northern Australia.

Spits and Sand Dunes

Coupled with beach and barrier beach coasts, there are various distinctive landforms. Spits, for instance, are common. A spit is created when a tongue of sand (and sometimes silt or even pebbles) is formed by waves and currents, so extending the beach into a small point that projects into a bay or the sea – the northern tip of Fraser Island, Queensland, is a good example.

A spit that connects a rocky offshore island to the mainland is called a tombolo – Barrenjoey, north of Palm Beach at the mouth of the Hawkesbury River, New South Wales, is one example. In some places spits may enclose bodies of water, so forming coastal lagoons.

Behind most beaches are sand dunes – massive amounts of sand that usually lie parallel to the shoreline with a depression (or swale) separating each successive dune. On Fraser Island, Queensland, - the world's largest sand island - the tallest sand dunes reach a height of 240m.

The dune nearest the coast, the fore-dune, often acts as a sand reservoir during storms and tempests, when beach sands may be eroded and deposited off-shore as bars. If stabilising dune vegetation is disturbed, then the sand is subjected to winds, which may form blow-outs. Extensive blow-outs may produce large areas of jumbled and drifting sand dunes called sand patches. Rabbits denuding the dunes' protective cover of vegetation formed the Delisser Sandhills at Eucla, Western Australia.

DID YOU KNOW?

Lake Macquarie is often cited as **Australia's largest lagoon**; it certainly is the largest singular lagoon body of water with one name.

Lake Macquarie, NSW

DID YOU KNOW?

The **longest elongated spit** or **silt jetty** is 9.7km; it is located where the Mitchell River enters Lake King, one of the interconnected Gippsland Lakes, Victoria.

Mt Tempest, on Moreton Island, Queensland, at a height of 280m, is considered to be the **highest sand dune in Australia.**

Fraser Island, and the adjacent mainland area of Cooloola, collectively known as the Great Sandy region, is the **greatest sand mass** along the Australian coastline. This region, along the sand islands of Moreton Bay, Queensland, traps sand carried by northward flowing coastal currents.

Fraser Island, Qld

Rock Coasts

Another type of coast is the rock coast, which is distinguished by low or high rocky cliffs. They may have small beaches at the base of the cliffs or lying between successive headlands. The beaches that lie between successive headlands may have distinguishing features similar to the beach coasts described above.

Some rock coasts, at least in their rocky sections, exhibit rock platforms – flat expanses of rock generally situated between high and low tide levels. Rock platforms are fascinating places to view the marine life that lives in or near rockpools, or has been left stranded by ebbing tides.

Rock coasts composed of readily eroding rocks also exhibit many erosional features such as stacks (isolated pillars of rock), blowholes, sea caves, sea gorges and sea arches. The Port Campbell coast of western Victoria has excellent examples of all of these features. Not all rock coasts exhibit rock platforms. In granite country sloping domes of granite rise from the sea. Such rock coasts are common along the Western Australian coast from east of Esperance to west of Albany, and also at Wilsons Promontory, Victoria. These coasts can be dangerous as king waves occasionally wash well up the domes' surfaces, whisking away unaware fisher folk and sightseers. It is interesting to note that the sloping granite dome of Rodondo Island in Bass Strait, visible from Wilsons Promontory, Victoria, rises 350m above sea level.

In places where rocky or undulating countryside has met with rising sea levels (as a result of the ending of the last ice age), drowned valleys or rias are common. Sydney Harbour and the mouth of the Hawkesbury River, New South Wales, as well as the Kimberley coast, Western Australia, are excellent examples of drowned valleys.

DID YOU KNOW?
On Tasman Peninsula, Tasmania, **the Tasman Arch, at 52m, is probably Australia's highest sea arch.**

Drowned Valley, Burrup Peninsula, WA

THE COAST

Table 6.6 Highest Sea Cliffs in each State/Territory

NSW	the sandstone cliffs near Garie, south of Sydney	120m
Vic	cliffs near The Twelve Apostles, Port Campbell district	60-100m
Qld	probably Indian Head, Fraser Island	120m
SA	Cape Torrens, Kangaroo Island	220m
WA	Zuytdorp Cliffs, south of Shark Bay	250m
Tas	Cape Pillar, Tasman Peninsula	275m
NT	cliffs on Marchinbar Island, Wessel Islands Group	70m
ACT	Steamers Head, Jervis Bay Territory	135m
OSA	below Mt Gower, Lord Howe Island	770m

Table 6.7 Some High Sea Cliffs and Near Vertical Coastal Hillslopes

eastern side of Mt Bowen, Hinchinbrook Island, Qld	1140m
below Mt Gower, Lord Howe Island, NSW	770m
Balls Pyramid, Lord Howe Island Group, NSW	558m
between Black Head and Resolution Point, south-east Tas	400m
the cliffs and hillslopes above Coalcliff, Illawarra, NSW	300m
between Cape Torrens and Harveys Return, Kangaroo Island, SA	260m
the hills north of Cape Jarvis, Fleurieu Peninsula, SA	180m
Rattlesnake Point, Cape York Peninsula, Qld	120m

The figures in table 6.6 represent sheer cliffs or near vertical hillslopes and may not be complete.

It is worth noting that the Zuytdorp Cliffs, Western Australia, are 130km in length; these cliffs are not as long as the Nullarbor sea cliffs in South Australia though, which extend for 210km. Some other high sea cliffs are presented in table 6.7.

DID YOU KNOW?

The cliffs at Cape Pillar, Tasmania, are the **highest sheer coastal cliffs of Australia** and extend for another 275m below sea level, giving a total fall of 550m. The sandstone cliffs at Steamers Head, Jervis Bay Territory, are the **highest sheer coastal cliffs on the Australian mainland**. Worthy of note is that Deal Island Lighthouse, Bass Strait, Tasmania, is **Australia's highest lighthouse: 305m above sea level.**

Table 6.8 Dramatic Coastlines in each State/Territory

NSW	Illawarra coast between Bundeena and Austinmer
Vic	Otway and Port Campbell coast between Aireys Inlet and west of Peterborough
Qld	North of Cairns coast between Ellis Beach and Port Douglas
SA	Nullarbor coast between Head of Bight and Border Village
WA	Kimberley coast between King Sound and Walcott Inlet
Tas	Tasman Peninsula coast between Haines Bight and Fortescue Bay
NT	Marchinbar Island south-east coast, Wessel Islands Group
ACT	Jervis Bay Territory coast between Cape St George and Governor Head

Rock coasts are known for their dramatic coastlines. This table lists what can be considered to be the most dramatic coastline of each state.

Glasshouse Rocks, NSW

Tidal Plain Coasts

Very flat coasts with vast tidal plains that are regularly or seasonal inundated are called tidal plain coasts. These are coasts where the land is forming at the expense of the sea. Found in areas of still waters, they are dominated by mudflats, salt marshes, mangrove formations and estuaries. Good examples can be seen at Corner Inlet, Victoria, the southern coast of the Gulf of Carpentaria, Queensland and the area around Derby, Western Australia.

Port Franklin jetties, Vic

Table 6.9 Some Long Australian Jetties

Lucinda Point Jetty, Qld	5760m
Cape Lambert Jetty, WA	2700m
Busselton Jetty, WA	1841m
Carnarvon Jetty, WA	1493m
Port Germain Jetty, SA	1283m
Beachport Jetty, SA	722m

DID YOU KNOW?

Tidal plains and other low coasts require the building of long jetties to reach deep water if goods are to be transferred onto ships. Originally wood was used to build these jetties – for example, Busselton, Carnarvon and Port Germain. Port Germain often claims to be **Australia's longest wooden jetty** – its original length was 1664m – but Busselton beats it. Modern long jetties are constructed of steel. Cape Lambert Jetty stands 17.87m high to allow for a high tidal fluctuation, while the Lucinda Point Jetty is so long that it actually dips to allow for the curvature of the Earth.

Coral Coasts

Along parts of the tropical and sub-tropical coast are found coral reefs. These coasts are very subdued and the reefs are under water at high tide. Reefs surrounding areas of still water form coral lagoons. Upon some reefs coral polyps have built up to form coral islands – where sand has collected they produce small islands called cays. Vegetated cays, especially if they have coconut palms, create the classic 'desert islands' found in movies and cartoons. Coral coasts are common off central and northern Queensland, forming the Great Barrier Reef, and the Exmouth Peninsula, Western Australia, forming Ningaloo Reef.

Table 6.10 Coral Reefs

Great Barrier Reef, Coral Sea	2300km
Ningaloo Reef, Indian Ocean	250km
Houtman Abrolhos, Indian Ocean	80km

The above figures represent total lengths. The Great Barrier Reef is in fact composed of up to 3400 individual reefs ranging in size from a few thousand square metres up to 120sq km; width of reefs range up to 250km.

DID YOU KNOW?

The Houtman Abrolhos reefs are the largest **most southerly coral reefs in the world**. Recent marine surveys have discovered a number of coral reefs in the Gulf of Carpentaria.

Islands

In the seas and oceans surrounding Australia are numerous islands. Most islands associated with the states and territories are continental islands. In other words, they were formed as a result of being separated from the mainland by rising sea levels after the last ice age – the islands themselves being the summits of ancient hills or mountains. Some of Queensland's islands are coral islands or cays.

All of these islands are found on the continental shelf, the undersea margin of the Australian continent, which varies in depth from 0 to 200m. The shelf varies in width from a few tens of kilometres to hundreds of kilometres. It connects the mainland to Tasmania and New Guinea and underlies vast bodies of water such as Bass Strait and the Gulf of Carpentaria. In some places the shelf is occupied by submarine canyons, deep gashes in the shelf that reach oceanic depths. The deepest ocean trench near the Australian mainland, the Diamantina, lies off Point D'Entrecasteaux, Western Australia, at a depth of 6100m.

Located well away from the Australian continent are oceanic islands that rise from the ocean floor. The islands are either volcanic in origin or are exposed portions of oceanic ridges. These islands are mostly Australian dependencies and include Norfolk, Lord Howe (part of New South Wales), Macquarie (part of Tasmania), Heard, McDonald, Cocos, and Christmas islands.

The water depths surrounding Christmas Island are awesome. The island rises above the Java Trench, the Indian Ocean's deepest point at 7,725m. The drop-off into the depths, beginning barely 20m from the shoreline, falls 500m within 200m.

Fitzroy Island, Qld

Table 6.11 Number of Australia's Islands

States/Territories	
NSW	102
VIC	184
QLD	1955
SA	346
WA	3747
TAS	1000
NT	887
ACT	1
Oceanic Islands	
Pacific Ocean	4
Southern Ocean	4
Indian Ocean	5

All different types of islands are included – estuarine, continental and oceanic islands, and coral formations. If just about everything from barely exposed mud and sand banks to rocky islets and stream-based islands are included, then Australia has approximately 12,000 islands.

DID YOU KNOW?

Australia has some pretty large islands. Australia itself is an island – considered to be the **world's largest island** and the world's **only island continent** (neither strictly true for the Eurasian-African continental land mass is also an island).

Hobart, Tas

Table 6.12 Australia's Largest Islands

Tasmania	67800sq km
Melville Island, Arafura Sea, NT	5786sq km
Kangaroo Island, Southern Ocean, SA	4416sq km
Groote Eylandt, Gulf of Carpentaria, NT	2285sq km
Bathurst Island, Arafura Sea, NT	1693sq km
Flinders Island, Bass Strait, Tas	1359sq km
Mornington Island, Gulf of Carpentaria, Qld	1002sq km
King Island, Bass Strait, Tas	829sq km
Dirk Hartog Island, Shark Bay, WA	618sq km
Hinchinbrook Island, Coral Sea, Qld	393sq km
Cape Barren Island, Bass Strait, Tas	335sq km

Table 6.13 Largest Island in each State/Territory

NSW	Broughton Island, Tasman Sea	4.6sq km
Vic	French Island, Westernport	167sq km
Qld	Mornington Island, Gulf of Carpentaria	880sq km
SA	Kangaroo Island, Southern Ocean	4416sq km
WA	Dirk Hartog Island, Shark Bay	618sq km
Tas	Flinders Island, Bass Strait	1359sq km
NT	Melville Island, Arafura Sea	5786sq km
ACT	Bowen Island, Jervis Bay	0.53sq km
AUS	Tasmania	67800sq km

Dirk Hartog Island, WA

Straits

Separating many islands from the mainland, or from each other, are bodies of water known as straits. A strait is a narrow stretch of sea connecting two broad expanses of sea. Sometimes straits are called passages or channels – for example the Backstairs Passage off Fleurieu Peninsula, South Australia, or the D'Entrecasteaux Channel off Bruny Island, Tasmania.

Table 6.14 Lengths of some Australian Straits

Bass Strait, off northern Tasmania	350km
Torres Strait, off northern Queensland	320km
Investigator Strait, off Yorke Peninsula, SA	125km
Apsley Strait, between Melville and Bathurst Islands, NT	75km
Great Sandy Strait, between the mainland and Fraser Island, Qld	75km
Whitsunday Passage, The Whitsundays, Qld	70km
Sunday Strait, off Dampier Land, WA	65km
Backstairs Passage, off Fleurieu Peninsula, SA	65km
D'Entrecasteaux Channel, between the mainland and Bruny Island, Tas	65km
Cadell Strait, off Arnhem Land, NT	60km
Dundas Strait, off Coburg Peninsula, NT	50km
Endeavour Strait, Cape York Peninsula, Qld	45km
Banks Strait, off north-east Tasmania, Tas	40km
Clarence Strait, between the mainland and Melville Island, NT	35km

The lengths in table 6.14, although approximate, are indicative of Australia's longest straits. Given the size of Bass Strait and Torres Strait, though they bare the name strait, their broadness would perhaps suggest that they might better be known as 'seas' – the Bass Sea and Torres Sea.

Fleurieu Peninsula, SA
(Left and right)

THE COAST

Tides

Washing Australia's coastline are tides. The alternate rising and fall of the sea – the tides – is caused by the gravitational pull of the moon and the sun.

The largest tides are the spring tides, which occur when the earth, sun and moon are in a line. The gravitational pull of the sun and moon are greatest at this time, making for higher high tides and lower low tides. When the earth is closest to the sun (around January), these effects are even greater. The converse of a spring tide – called a neap tide – occurs when the sun and moon are at right angles to each other. Then, they partly cancel out each other's tidal influence, which results in lower high tides and higher low tides. Spring and neap tides occur twice a month.

Tides usually occur twice a day, but south of Geraldton, Western Australia, there is only one tide (high or low) a day. At Hamelin Pool in Shark Bay, Western Australia, the tides are barely discernable.

Table 6.15 Tides

Spring Tidal Ranges

Least
- south of Perth, W, — up to 60cms

Greatest
- Collier Bay, The Kimberley, WA — 10.97m
- Derby, The Kimberley, WA — 10.36m
- Broome, The Kimberley, WA — 8.53m
- Port Darwin, Top End, NT — 7.31m
- different sites in Broadsound, Central Coast, Qld — 7.31 to 9.14m
- Wyndham, The Kimberley, WA — 7.01m

Longest Tidal Bore
- up the Victoria River, NT — 80km

Tidal Races
- The Rip, Port Phillip Heads, Vic — up to 14.4km/hr
- Wyndham, The Kimberley, WA — up to 17km/hr
- Walcott Inlet, The Kimberley, WA — up to 28km/hr

Table 6.16 Spring Tidal Ranges in the Capital Cities

NSW	Fort Denison, Sydney	1.55m
Vic	Port Phillip Heads, Melbourne	1.60m
Qld	Brisbane Bar, Brisbane	1.98m
SA	Port Adelaide, Adelaide	2.51m
WA	Fremantle, Perth	0.76m
Tas	Derwent River, Hobart	1.37m
NT	Port Darwin, Darwin	7.31m

The figures in table 6.16 represent average high and low-water spring tidal ranges.

Tidal ranges vary for many reasons, including shallowness of water, constrictions in estuaries, and so on. The most interesting tidal activity happens in Western Australia where the greatest and least tidal ranges are located. Where tidal ranges are great, long piers must be built for ships to anchor at low water. It is not uncommon to see small boats sitting on the bottom at low tide in these places.

Rock coasts with great tidal ranges can exhibit tidal waterfalls – for instance, the Horizontal Falls at Talbot Bay in the Kimberley region of Western Australia. The ebbing and flooding tides entering or leaving constricted inlets have different water levels, thus producing a waterfall.

Tidal bores are wave-like features, in which the oversteep wave of the tidal front produces a surf. Tidal race indicates the speed of a tide. In deep water this is approximately 1km/hr and in shallow water, up to 7km/hr; these speeds being faster in constricted areas.

Row boat, ShuteHarbour, Qld

CHAPTER SEVEN
THE BUSH

THE BUSH

Classification of Vegetation

The vegetated countryside, on first appearance, presents a bewildering array of living matter. Its life forms vary from district to district - sometimes dramatically, more often subtly. Depending on rock type, topography, various weather phenomena, fires, soils, and the influence of animal and human activity, vegetation produces different types of what are called formations.

Nevertheless, the many different types of vegetation formations do exhibit a general trend towards uniformity in their appearance. Consequently, to a traveller in a speeding car the vegetation appears as a hodge-podge of trunks, leaves, shrubs and grasses - seemingly without form, but the countryside does not have to be perceived in this way.

It is possible to view vegetation systematically. Although no classification is perfect, owing to the complexity of variables involved, one particular formation can be differentiated from another by simply looking at how each one is composed. One common system of differentiation used in Australia is based on the dominant type of vegetation (that is, tree, shrub or herb), its height above ground, and the 'ground area' covered by the foliage, which is known as its projective foliage cover.

With the use of this classification system, vegetation can be grouped into various formations, such as different types of forest, woodland, scrub, heath, shrubland and herbland. Other formations do not fit so readily into this classification because they occupy extreme habitats such as tidal flats, wetlands, or sand dunes. The most common of these other types include mangrove formations, the incredible diversity of wetlands, the plant life on coastal sand dunes and, offshore, seagrass formations.

Overall, there are an estimated 20,000 species of native plants in Australia. Of these, about 17% are considered rare or threatened with extinction. Almost half of the native plants under threat are found in the south-western portion of Western Australia, the area of Australia most thoroughly cleared of native vegetation.

The south-east and south-west corners of the continent are interesting places for rare and endangered euclypts. Many different types are found there, occupying small niches in stands that can be measured in tens or hundreds of plants. Many are found in cold or exposed locations. Not only are some of them relics of past cooler climatic conditions, but these plants may well become the dominant plants in future vegetation formations when the planet once again swings towards another ice age - a good reason for preserving our endangered species.

DID YOU KNOW?

The Mongarlowe mallee (*Eucalyptus recurva*) is probably **Australia's most endangered plant**, there being barely five or six living specimens anywhere in existence. They are found near Mongarlowe, Southern Tablelands, New South Wales.

Acacia

Table 7.1 Species Numbers of Common Australian Woody Plant Types

Acacia (wattles, mulga etc)	700+
Eucalyptus (gum trees, mallees etc)	440
Grevillea	250
Hakea	140
Melaleuca (paperbarks)	140
Eremophila (emu bushes)	90
Banksia	80
Casuarina and *Allocasuarina* (she-oaks)	40+
Cassia	40
Atriplex (saltbushes)	40
Maireana (bluebushes)	40
Terminalia (nutwoods)	30

Eucalyptus

Table 7.2 Species Numbers of Common Australian Herbal and Other Plant Types

Liverworts	1000
Asteraceae (daisies)	800+
Poaceae (grasses)	700+
Orchidaceae (orchids)	600+
Mosses	600

The figures given for species numbers in tables 7.1 and 7.2 are approximate. It should be noted, too, that the genus *Eucalyptus* has recently been broken up into a number of different genera.

Table 7.3 Growth Rates of Trees and other Plants

Stromatolites, WA	50mm per century
Waddywood (*Acacia peuce*), Qld-NT	150mm per century
Grass tree (*Xanthorrhoea australis*), WA	250mm per century
Grass tree (*Kingia australis*), WA	300mm per century
Huon pine (*Lagarostrobos franklinii*), Tas	1200mm per century

Stromatolites are not strictly plants in the normal sense of the word but rather a type of blue-green algae known as cyanobacteria. Blue-green algae are the world's oldest known form of life, having existed on the planet for the last 3500 million years. Matted sheets of this algae trap sediments that slowly grow and build up into what is known as stromatolites. They exist in a few colonies around the world, the best examples being at Hamelin Pool, Shark Bay, Western Australia.

The growth rate given in table 7.3 was derived from a single source so it may not be reliable, as waddywood seedlings are known to grow much faster than this. Given that waddywood grows to a height of nearly 15m, then at this growth rate mature specimens would be about 10,000 years old, a very improbable age.

Grass Tree, Jingalup, WA

Table 7.4 Ages of Trees and other Plants

Huon pine (***Lagarostrobos franklinii***), Tas	2500 years
Antarctic beech (***Nothofagus moorei***), NSW-Vic-Tas	2000 years
Pencil pines (***Athrotaxis cupressoides***), Tas	2000 years
Cycads (various species), various States/Territories	1500 years
Kauri pine (***Agathus robusta***), Qld	1100 years
Karri (***Eucalyptus diversicolor***), WA	1000 years
Grass tree (***Kingia australis***), WA	1000 years
River red gum (***Eucalyptus camaldulensis***), various states	1000 years
Jarrah (***Eucalyptus marginata***), WA	1000 years
Baobab (***Adansonia gibbosa***), WA-NT	1000 years
Tree ferns (***Cyanthea/Dicksonia spp.***), NSW-Vic-Qld-Tas	500 years

The figures in table 7.4 represent typical ages of mature species.

A zamia palm at Mt Tambourine, McPherson Range, Queensland, now destroyed, was once claimed to be over 1500 years old. The Prison Tree, near Wyndham, Western Australia, is claimed to be 4000 years old. There has also been a claim made that there is a 10,000 year old Huon pine at the Lake Johnson Nature Reserve, in Tasmania.

DID YOU KNOW?

Planted in 1816, at the Botanic Gardens, Sydney, New South Wales, a number of swamp mahoganies (*Eucalyptus robusta*) are **Australia's oldest street trees**.

Mountain ash tall open forest, Gunyah,

Kangaroo paws, open forest understorey, Perth,

Australian Vegetation Types
Closed forests

Traditionally called jungles, closed forests are normally known as rainforests or, in some places, scrubs. Such forests are found in the well-watered parts of eastern coastal Australia, Tasmania and in some isolated pockets along the northern coastline. The many different types of closed forest range from tropical rainforests such as those in the Wet Tropics of far north Queensland, to cool temperate rainforests found in Tasmania and on the higher coastal ranges of south-east Australia. Elsewhere there are the so-called dry rainforests, found mostly in drier and more inland localities; there are monsoonal rainforests, too, which occupy small niches around the billabongs or within the gorges of northernmost Australia. All but the cool temperature rainforests generally have a rich assemblage of plant life. Rainforests are not fire tolerant.

Cleared salmon gum woodland, Wyalkatc

THE BUSH

Table 7.5 Australia's Tallest Tree Species

Species	Distribution	Max Height
Mountain ash (*Eucalyptus regnans*)	Vic-Tas	100m+
Alpine ash (*E. delegatensis*)	NSW-Vic-Tas	90m
Shining gum (*E. nitens*)	NSW-Vic	90m
Messmate stringybark (*E. obliqua*)	NSW-Vic-Qld-SA-Tas	90m
Karri (*E. diversicolor*)	WA	85m
Flooded gum (*E. grandis*)	NSW-Qld	75m
Maidens gum (*E. maidenii*)	NSW-Vic	75m
Red tingle (*E. jacksoni*)	WA	70m
Tallowwood (*E. microcorys*)	NSW-Qld	70m
Blackbutt (*E. pilularis*)	NSW-Qld	70m
Tasmanian blue gum (*E. globulus*)	Vic-Tas	70m
Red cedar (*Toona australis*)	NSW-Qld	70m
Spotted gum (*Eucalyptus maculata*)	NSW-Qld	70m
Sydney blue gum (*E. saligna*)	NSW-Qld	65m
Round-leaved gum (*E. deanei*)	NSW-Qld	65m
Mountain grey gum (*E. cypellocarpa*)	NSW-Vic	65m
Hoop pine (*Araucaria cunninghamii*)	NSW-Qld	60m
Norfolk Island pine (*A. heterophylla*)	Norfolk Island	60m

Open forests

Open forests are located in the moist, coastal districts ranging, with minor exceptions, from The Kimberley in Western Australia, across northernmost Australia, right down the east coast and across the southern coast of Victoria, as well as in Tasmania, south-western Western Australia, and along parts of the Mt Lofty Ranges and southern Flinders Ranges in South Australia. Most open forests are dominated by eucalypts although acacias (especially brigalow), casuarinas, cypress pines and paperbarks also form open forests. It is the eucalypt open forests that are most familiar to the capital city dwellers, for each city has eucalypt open forests on its doorstep.

Two different types of eucalypt open forest can be readily distinguished. Tall open forests, sometimes referred to as wet sclerophyll forests, occupy moist gullies and southerly aspected hillslopes in moist habitats. They include Australia's tallest trees and many are harvested for timber and woodchips. The second type of forest is Australia's most common forest type and is also called 'open forest' or, sometimes, dry sclerophyll forest. This type of forest is distinguished by trees with a maximum height range of 10–30m. The understorey of such forests are sometimes shrubby (such as the ones seen around Sydney, New South Wales, or Perth, Western Australia), or grassy (common in Queensland).

Many of these eucalypt forests are referred to as old growth forests - that is, they are ecologically mature and have been subjected to minimal disturbance. These forests have high conservation values as well as being particularly beautiful to behold.

Table 7.5 shows the maximum representative heights for trees growing under ideal conditions. All the eucalypts are found in tall open forests (wet sclerophyll forests); the tallowwood and shining gum may be located on the edge of some rainforests while hoop pines and red cedars are found in rainforests.

The baobab (*Adansonia gibbosa*) is a tree that consistently has a very large girth – up to 20m has been measured on a tree near the junction of the Sprigg and Isdell rivers in the Kimberley, Western Australia. A red tingle tree (*Eucalyptus jacksoni*), near Walpole, Western Australia, has a hollowed out buttressed girth of 22.3m.

Snowy Mountains

DID YOU KNOW?

By way of interest, **the trees that grow at the highest altitude in Australia** are snow gums (*Eucalyptus pauciflora*) found in the vicinity of the upper Snowy River-Mt Kosciuszko area, Snowy Mountains, NSW. At around 2000m these attractive plants form a treeline between the sub-alpine and alpine zones. This treeline is at lower elevations in sub-alpine valleys, as well as in Victoria and Tasmania.

Table 7.6 Some Tall Australian Trees

Tree	Height
Mount Tree, mountain ash, Styx Valley, Tas	96.5m
Damocles Tree, mountain ash, Styx Valley, Tas	95.2m
Mr Jessop Tree, mountain ash, Wallaby Creek, Vic	91m
Maydena Tree, mountain ash, Styx Valley, Tas	89m
Stewarts Karri, karri, near Manjimup, WA	88m
Andromeda Stand, mountains ashes, Styx Valley, Tas	up to 88m
Grandis Tree, flooded gum, near Bulahdelah, NSW	87m
Geeveston Tree, mountain ash, Tas	87m
Giant Stringybark, Beech Creek area, Tas	86m
Karri, near Warren River, WA	84.7m
The Big Tree, mountain ash, Cumberland Valley, Vic	84m
Wayatinah Tree, mountain ash, Tas	84m
Gandalfs Staff, mountain ash, Styx Valley, Tas	84m
Chapel Tree, mountain ash, Styx Valley, Tas	83m
Noble Tree, white gum, near Styx River, NSW	79.2m
El Grande, mountain ash, Styx Valley, Tas	79m
Four Aces (4 karris), near Manjimup, WA	up to 79m
Christmas Tree, mountain ash, Styx Valley, Tas	77m
Ada Tree, mountain ash, near Powelltown, Vic	76m
Dale Evans Bicentennial Tree, karri, near Pemberton, WA	70m
Bird Tree, blackbutt, near Kendall, NSW	69m
Big Fella Gum Tree, flooded gum, near Kendall, NSW	67m
Gloucester Tree, karri, near Pemberton, WA	60m
Diamond Tree, karri, near Manjimup, WA	51m
Boorara Tree, karri, near Northcliffe, WA	50m
Kalpowar Queen, hoop pine, near Monto, Qld	46m

The figures in table 7.6 are not complete – there are quite a number of mountain ashes in Victoria and Tasmania between 75 and 90m tall. Nor are tree heights constant – wind damage, lightning strikes and old age mean that the figures are not always reliable.

In New South Wales, the Grandis Tree's figure includes a dead branch at its top; its living section is 84.3m tall. The Big Tree, Victoria, if in fact it is the same tree once known as the Cumberland Tree, has been measured at nearly 92m tall, with a girth of 4.1m measured about 3m above the ground. Nearby there were once 25 other tall trees with an average height of 81m. The Cornwaithe Tree, a mountain ash felled at Thorpdale, Victoria, at 114.3m tall, was probably the tallest in Victoria. It has been claimed that a mountain ash felled near Healesville, Victoria, was 132.6m tall, or a total of 145m when the topmost branches and stumped were added – this claim is unreliable and not officially recognised. Also unreliable is the claim for the Ferguson Tree, chopped down at Watts River in Victoria, and claimed to be 152m high.

The Gloucester Tree, Western Australia, was originally 76.2m tall before a bushfire lookout was built near its top. There are records of a karri felled near Pemberton, Western Australia, being 87m tall. The tallest tree in South Australia, located on private property, near Mt Gambier is probably a 50m high river red gum – it is estimated to be 800 years old.

Gloucester tree

DID YOU KNOW?

The Maydena Tree, Styx Valley, Tasmania was once Australia's tallest tree, at 98.75m, before a lightning strike lopped off its top. This tree is approximately 61m to its first branch, with a girth of 12.8m, and an estimated age of 300 – 350 years. The El Grande Tree, once **Australia's bulkiest tree** at 404 cubic metres, was unfortunately killed by a Forestry Tasmania supervised regeneration fire in April, 2003. The Geeveston tree, with a bulk of 210 cubic metres, is possibly **Australia's largest eucalypt**.

DID YOU KNOW?

By way of contrast, the **shortest plant** would appear to be the duckweed *Wolffia angusta*, found in tropical Australia, which grows to a height of 0.6 millimetres!

Woodlands

Of all Australia's treed formations, woodlands are the most common type. The many varieties that exist are categorised according to dominant species, dominant tree heights and types of understoreys, which may be grassy, shrubby, or with spinifex. Woodlands are generally located on the drier, interior side of the open forests, though in some localities they are found about the coast. Like open forests, woodlands are dominated by eucalypts, acacias, casuarinas, cypress pines and paperbarks, with eucalypts and acacias being the most common; many other species are well represented too. The eucalypt woodlands with an understorey of grass – the so-called grassy or savanna woodlands – are the most common type, being prevalent throughout northern Australia and on the inland side of the Great Dividing Range, where they have been extensively cleared for grazing and cropping.

DID YOU KNOW?

The **most widespread eucalypt tree** is the river red gum (*E. camaldulensis*), which is found forming riverine forests and woodlands in all mainland Australian states. Probably **Australia's most common eucalypt** is the variable barked-bloodwood (*E. dichromophloia*). Its range covers most of the semi-arid regions of northern Australia.

THE BUSH

Scrubs and heaths

There are both open and closed scrubs and heaths. Closed scrubs (which should not be confused with rainforest scrubs) are distinguished by densely growing shrubs – typically tea-trees or paperbarks – occupying swampy areas or depressions behind coastal dunes.

Open and closed heaths are not widespread in Australia but, owing to their location, are familiar to many city and coastal dwellers. They occupy sandy soils in coastal areas or coastal headlands, as well as some sand plains and sandstone plateaus. Australian heaths are relatively dense, scratchy formations, often of great diversity, and they like a good bushfire from time to time. In fact, many heath-formation plants need fire to germinate. While heath formations without trees are generally uncommon, many eucalypt open forest and woodland formations have a shrubby understorey composed of heathy plants. The sandy soils of the dry regions of southern Australia, abutting the semi-arid and arid areas, support quite large tracts of open heaths – for instance, the Ninety-Mile Desert of South Australia and the Big Desert of Victoria.

Open scrubs are like the open forests of the shrub world and most of them are dominated by the mallee. A kind of eucalypt, the mallee is a multi-stemmed woody plant (a shrub) up to 8m tall and comes in two basic types: the bull mallee, which has just a few thick stems; and the whipstick mallee, which has many slender stems. The understorey varies from shrubby to grassy and spinifex. Open scrubs are mostly located in southern Australia: ranging from western Victoria to north-western Western Australia, lying on the coastal side of the semi-arid regions; with minor areas elsewhere, usually in cold climates or in areas of impoverished soils. Much of the Australian mallee country has now been cleared for agriculture.

The *tallest heath in Australia*, *Richea pandanifolia*, found in Tasmania, grows up to 9m in height.

Shrublands

In terms of the area occupied, shrublands are perhaps the most widespread of Australia's vegetation formations. Some shrublands are dominated by mallee eucalypts (being less dense than their open scrub counterparts) but most are dominated by acacias, mulga (*Acacia aneura*) being the commonest. Mulga shrublands are found throughout much of semi-arid inland Australia. While only relatively small areas have been cleared, these formations have been subjected to extensive grazing of sheep and cattle on vast outback stations. Many of Australia's sand dune deserts, too, support sparse tracts of open shrublands dominated by acacias.

Low shrublands are mostly dominated by short stature shrubs (up to 2m, and normally less than 1m) such as saltbushes and bluebushes. They are widespread across semi-arid and arid southern Australia, forming vast, empty plains or shrub-steppes like the Nullarbor Plain or the Hay Plain of New South Wales. Most low shrublands are extensively grazed by sheep and have suffered serious depletion and erosion problems in the past.

DID YOU KNOW?

Probably the **most common Australian woody plant** is the mulga shrub (*Acacia aneura*). It sometimes takes a treed form; found in a number of vegetation formations, especially shrublands, it covers an estimated area of around 2.5Msq km.

Mulga shrublands, near Thargomindah, Qld

Table 7.7 Some Tall Shrub Species

Species	Height
River cooba (*Acacia stenophylla*), mainland states	15m
Yarran (*A. homalophylla*), NSW, Qld	12m
Cooba (*A. salicina*), mainland states	12m
White mallee (*Eucalyptus dumosa*), NSW, Vic, SA	12m
Mulga (*Acacia aneura*), mainland states (not Vic)	10m
Green mallee (*E. viridis*), NSW, Vic, Qld, SA	10m

The figures in table 7.7 are maximum heights grown under ideal conditions. Shrubs that are as tall as this may take on a treed form, though still displaying the appearance of a shrub, with many trunks and branches growing from a singular trunk near ground level.

Closed heath, north of Esperance, WA

Herblands

There are many types of closed herblands, dense formations of non-woody plants, such as alpine herbfields (found near and above the treeline), tussock grasslands (found on basalt soils or in alpine areas), sedgelands (for instance, the button-grass plains of south-west Tasmania), and fernlands. However, closed herblands occupy only small areas.

There are also herblands, which are far more widespread and common than closed herblands – though mostly in inland semi-arid and arid Australia. Tussock grasslands dominated by Mitchell grass are the most common type of herbland, being found across parts of northern Australia and western Queensland. These Mitchell grass plains produce the 'big-sky country' which many Australians think is typical of the outback: it is, but only in some places. Vast horizons, mirages and shadowy jump-ups (low flat-top hills) typify this extensively grazed country.

Other herblands occupy the coastal plains of northern Australia and produce distinctive formations, but particular mention should also be made of a unique Australian herb – the spiky grass commonly known as spinifex or porcupine grass. Spinifex rarely forms formations in its own right but as an understorey plant it occupies about one-fifth of the continent, mostly in the arid regions. Many of Australia's deserts, while not that dry by world standards, are deemed deserts because of the unpalatability of spinifex to grazing stock.

Alpine closed herbland, NSW

DID YOU KNOW?

Possibly **Australia's largest herb**, a rough tree fern (*Cyathea australis*) located in the Mt Dromedary Forest Preserve, New South Wales, has been measured as being 9.5m tall.

Wetlands

Wetlands are tracts of countryside subjected to permanent, seasonal or irregular inundation and can be one of two basic types: freshwater or marine. Wetlands, once despised as marshes and swamps, are important aquatic habitats supporting a vast range of plant and animal life. In fact, they are normally distinguished by the vegetation they support; in this respect they are different to lakes.

Numerous types of freshwater wetlands are found throughout Australia, especially in the humid and sub-humid country of eastern and south-western Australia or in the monsoonal north. Examples of freshwater wetlands include: river and river flats; creeks and creek margins; various herb, sedge, lignum and canegrass swamps; button-grass plains; reed and shrub marshes; wet heaths; open bodies of water containing aquatic plants.

Marine wetlands, being coastal wetlands, include such types as tidal estuaries, mangrove formations (there are freshwater mangroves as well), coastal lagoons and salt marshes. Their defining feature is salty or brackish water, which obviously influences the types of vegetation found in these wetlands.

Alpine tussock grassland, Mt Hotham, Vic

Freshwater wetland, Bamarang, NSW

Table 7.8 Some Australian Wetlands

Morton Bay Wetlands, Qld	3000sq km
Kakadu Wetlands, NT	2330sq km
Daly-Reynolds Floodplain, NT	1590sq km
Adelaide River Floodplain, NT	1350sq km
Mary River Floodplain, NT	1280sq km
Wooramel Seagrass Bank, WA	1000sq km
Macquarie Marshes, NSW	950sq km (1900sq km)
Jardine River Wetlands, Qld	820sq km
Arafura Swamp, NT	710sq km
Port Darwin Wetlands, NT	480sq km
Burdekin Delta, Qld	320sq km
Lowbidgee Floodplain, Murrumbidgee River, NSW	200sq km (400sq km)

In table 7.8 the figures in brackets are the original sizes of those wetlands before human interference. Extraction for irrigation farming has led to significant reductions in the streams that feed these wetlands. Other well-known wetlands include the following: Great Cumbungi Swamp, Lachlan River, New South Wales (160sq km); Gwydir River Wetlands, New South Wales (50sq km - 270sq km original size); Noosa River Wetlands, Queensland (99q km); Lower Daintree River Wetlands, Queensland (53sq km); Coongie Lakes, South Australia (77sq km); Bool-Hack Lagoons, South Australia (36q km).

Changes to Vegetation Formations

The most significant changes to Australia's natural vegetation formations – and consequently, to the Australian countryside – have been caused by the clearing of the land for grazing, agriculture, mining and urbanisation, and by the intentional use of, or incidental damage produced by bushfires.

Fire in itself will not drastically alter the vegetation formations in the long term, where those formations have evolved with fire. In fact, many plants depend on occasional fires in order to regenerate. The problem is that changes in the intensity and frequency of fires does alter and damage formations, especially those ones not able to cope with fires (for instance, rainforests). The prevention of fires also changes formations, for in the absence of fire the growth of shrubs is promoted at the expense of grasses. The eucalypt woodlands dominated by bimble box within the sub-humid/semi-arid country of eastern Australia were once considered to be good grazing land by the early pioneers. The prodigious growth of shrubs in their understoreys has rendered many of these woodlands partly useless for such purposes however, and the use of fire as a management tool could have assisted in keeping this country open.

For hundreds of generations Aborigines had engaged in what is known as 'firestick farming', which had kept the countryside open and clear. By firing the land they promoted the growth

Water lilly wetland, St Lawrence, Qld

of grasses as well as flushing out game. Regulating their burning to small areas and firing only during the cool seasons, they avoided major conflagrations and produced a mosaic of recently burnt areas, regenerating areas and unburnt areas. This diversity enabled both plant and animal species to survive.

The partial or wholesale land-clearing that took place following European occupation had a far more dramatic impact. Where the removal of most or all of the vegetation had taken place, it was only a generation or two before there was an increase in soil erosion caused by the soil's greater exposure to winds and water. Streams became cloudy with sediment and waterholes silted up.

Maintaining stands of vegetation is important in the control of soil erosion. In humid country the crowns of trees provides protection for the understorey, which in turn protects a ground layer of grasses, mosses and fallen leaves. This ground layer absorbs rainwater, releasing it gradually and thus regulating surface runoff and stream flows, which mitigates the effects of flooding rains in extreme circumstances. In dry areas, stands of vegetation maintain a good ground surface mulch of humus as well as breaking the velocity of the wind, thereby minimising the effects of wind erosion and the desiccating effect of hot northerly winds.

On some cleared land salt has entered the upper levels of the soil. As trees were removed the water table rose, bringing with it salts deposited at deeper levels. Stands of trees can reduce this soil contamination by salt; by utilising the groundwater, the trees keep the water table below the root zone of grasses and shrubs. The net result of wholesale land clearances, then, was not only soil loss but also the loss of otherwise productive farmlands.

Stocking, too, has affected the countryside: the natural herbage has not only been replaced with exotic plants in the form of improved pastures but it has also been depleted by grazing and browsing. Furthermore, trampling and compaction of soil has occurred with the introduction of hard-hoofed animals. Not all these changes were intentional. For instance, the introduction of the rabbit, for hunting in the 1850s, has led to major changes in the Australian countryside, especially in the southern semi-arid and arid regions. Not content with replacing native animals such as the bilby, rabbits have severely depleted the vegetation of the rangelands, permanently altering the nature of the landscape.

The net result of all of these impacts is that the countryside has significantly changed since European occupation, in many cases for the worse. Unfortunately, massive land clearances are still occurring in Australia. Between 1983 and 1993 half a million hectares per year were cleared in Queensland alone, which accounted for 60 per cent of Australia's total (this figure includes the clearance of shrubby regrowth on already cleared land). This clearance rate is the equivalent of 70 football fields per hour and gave Australia the dubious ranking of the eighth-worst land-clearing country in the world for this ten-year period. Fortunately though, there is a growing awareness in the community that the land can be abused only for so long.

Saltbush low shrubland, near Ivanhoe, NSW

THE BUSH

Table 7.9 Australia's Vegetation Formations: Pre-1788

Formation	Area occupied	Typical plant types	% Cleared
Shrublands	3.07Msq km	mulga, other acacias	2%
Woodlands	2.18Msq km	mainly eucalypts	25%
Herblands	0.59Msq km	Mitchell grass	10%
Open scrubs	0.59Msq km	mallee eucalypts	35%
Low shrublands	0.46Msq km	saltbushes, bluebushes	2%
Open forests	0.40Msq km	mainly eucalypts	30%
Open herblands	0.12Msq km	Mitchell grass	1%
Open heaths	0.11Msq km	numerous plants	45%
Rainforests	0.07Msq km	numerous types	30%
Unvegetated areas	0.05Msq km	saltlakes, etc	0%
Seagrasses meadows	0.05Msq km	seagrasses	0%
Beach dune formations	0.03Msq km	various types	1%
Salt marshes	0.01Msq km	various types	2%
Mangroves	0.01Msq km	mangroves	4%

Table 7.9 lists the different types of vegetation formations, and the area they occupied, that existed before European settlement in 1788. The figures are approximate and are measured in millions of square kilometres. The land clearing figures, expressed as approximate percentages of the formation cleared, represent the situation around 2001. Prior to European occupation many vegetation formations were modified by Aboriginal 'firestick farming' practices, and of course this pattern of land management would have continued in the early years of occupation in areas where Aborigines still had control over their land.

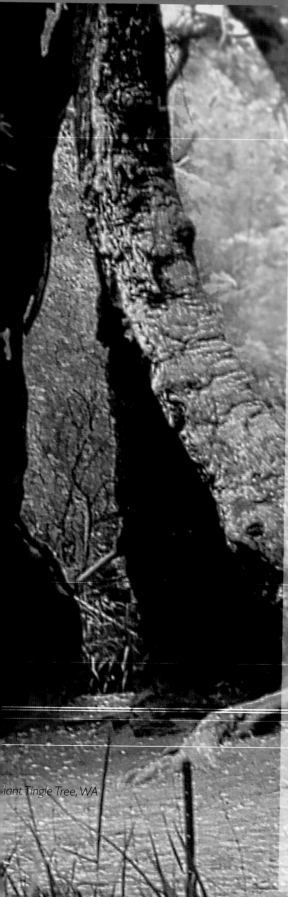

Giant Tingle Tree, WA

DID YOU KNOW?
The Giant Tingle Tree is regarded as being one of the **10 largest living things on the planet**.

Vegetation – the Human Dimension
The stumps of giant trees, after felling, were sometimes were put to good use during the pioneering days. Near Neerim, Gippsland, Victoria, a hollow stump was used as a house; elsewhere in Gippsland one tree stump had a track passing through it. Another tree stump in the same district, with an internal diameter of nearly 8m, was roofed and used over the years as a dairy, church, school and horse stables. It could accommodate 20 people.

Aboriginal people used trees for shelter, too. North of Melrose, South Australia, there are many tree shelters utilising hollows burnt out or enlarged by fires at the base of the trees' trunks.

Some Well-Known and Less Famous Australian Trees
Abbot Street Giant Fig – (*Ficus infectoris*), Cairns, Queensland. This big tree, now removed, had a foliage circumference of about 52 metres. It stood 21 metres high and had a girth of 6.8 metres.

Ancient Empire – red tingle trees (*Eucalyptus jacksoni*), near Walpole, Western Australia. A collection of red tingles that have girths up to 16m.

Beadells Tree – a gum tree marked in 1958 by Len Beadell (considered to be the last of the Australian explorers) located north-west of Warburton, Western Australia. Another 'Beadell Tree' is located near Warakurna, Western Australia.

Bilsons Tree – a river red gum (*Eucalyptus camaldulensis*), Casterton, Victoria. At 845 cubic metres this 800 year old tree is the biggest river red gum in Australia.

Cathedral Fig – a strangler fig (*Ficus destruens*) near Lake Tinaroo, Queensland. A striking fig set deep within a rainforest.

Curtain Fig – a banyan (*Ficus columnaris*) near Yungaburra, Queensland. Aerial roots up to 18m tall descend from its sloping trunk to form an impressive and photographic curtain.

Cazneaux Tree – a river red gum (*Eucalyptus camaldulensis*) near Wilpena Pound, Flinders Ranges, South Australia. An impressive tree estimated to be over 300 years old; made famous by the Cazneaux photograph 'Spirit of Endurance'.

Colindale Giant – a fig (*Ficus sp*), Colindale Station, north of Hughenden, Queensland. A large, spreading fig tree, it has a foliage circumference of about 115 metres and a girth of 6.7m, and is approximately 30m high.

Diamond Tree – a karri (*Eucalyptus diversicolor*), Manjimup district, WA. A former lookout tower standing on top of a 51m karri tree.

Dig Tree – a coolabah (*Eucalyptus microtheca*) at Cooper Creek, Queensland. The tree carved by members of the ill-fated Burke and Wills expedition.

Explorers Tree – a blackbutt (*Eucalyptus pilularis*) near Katoomba, New South Wales. The tree said to have been carved by Lawson; he, with Blaxland and Wentworth were supposedly the first Europeans to cross the Blue Mountains.

Foundation Tree – a South Australian blue gum (*Eucalyptus leucoxylon*), Glenelg, Adelaide, SA. The tree beneath which the government of the Colony of South Australia was proclaimed.

Four Aces – karris (*Eucalyptus diversicolor*), Manjimup district, Western Australia. Four tall karri trees ranging in height from 67m to 79m, and standing in near-perfect symmetry.

Giant Tingle Tree – a red tingle (*Eucalyptus jacksoni*), near Walpole, Western Australia. A giant tree with a girth of 25m.

Gloucester Tree – karri (*Eucalyptus diversicolor*), Pemberton, Western Australia. A famous tourist tree, complete with its terrifying spiral ladder giving access to a bushfire lookout on its top.

Grandis Tree – a flooded gum (*Eucalyptus grandis*) near Bulahdelah, New South Wales. A popular tourist attraction and the tallest tree in New South Wales.

Gregory Tree – a baobab (*Adansonia gibbosa*) near Timber Creek, Northern Territory. A tree marked by the explorer Gregory.

Herbig Tree – a red gum (*Eucalyptus sp*), Springton, South Australia. The temporary home of the Herbig family (Friedrich and Caroline), who lived in its hollowed trunk for two years in the 1850s and had two of their many children there.

Karri-with-a-Hole – a karri (*Eucalyptus diversicolor*) near Pemberton, Western Australia. A 76m high karri tree (before lightning lopped its top) with a hole through its trunk.

THE BUSH

Herbig Tree (left)

Prison Tree (left)

King Jarrah Tree – a jarrah (*Eucalyptus marginata*) near Manjimup, Western Australia. A superb specimen of a jarrah, said to be over 600 years old.
Largest Baobab in Captivity - (*Adansonia gregorii*), Wyndham, Western Australia. Its foliage circumference is not known; but, given its foliage diameter of 25m, a rough circumference would be about 75m.
Marianne North Tree – a karri (*Eucalyptus diversicolor*) near Pemberton, Western Australia. A tree made famous by the artist Marianne North.
Monkira Monster - a coolabah (*Eucalyptus microtheca*), Neurragully Waterhole, Channel Country, Queensland. Early reports suggest that with a foliage circumference of about 230 metres, it may have been Australia's most spreading tree. The tree was about 18 metres high, had a girth of 14m and branches reaching out about 36m.
Pekina Creek River Red Gum – (*Eucalyptus camaldulensis*), Orroroo, South Australia. One does not normally associate big trees with South Australia but this river red gum does have a trunk diameter of 3m and a foliage diameter of 20m.
Prison Tree – a baobab (*Adansonia gregorii*), Derby, Western Australia. A large-girthed (over 16m) and hollowed baobab used for holding up to 12 prisoners in days gone by. Another baobab prison tree in Western Australia is located on the King River near Wyndham; with a girth of 10m, it accommodated up to 18 prisoners. According to one source, the King River tree is 4000 years old!
Sturt Tree – a river red gum (*Eucalyptus camaldulensis*), Wentworth, New South Wales. The tree nearby where explorer Sturt weighed anchor in 1830 after the discovery of the junction of the Murray and Darling Rivers.
Tree of Knowledge – a ghost gum (*Eucalyptus papuana*), Barcaldine, Queensland. The birthplace of the Australian Labor Party.
Twin Ghost Gums – (*Eucalyptus papuana*), near Alice Springs, Northern Territory. These photogenic trees in the Macdonnell Ranges were a popular tourist attraction before their demise.
Twin Kauris – kauri pines (*Agathus robusta*), Lake Barrine, Atherton Tableland, Queensland. A pair of impressive native pines overlooking Lake Barrine.
unnamed fig – (*Ficus sp.*), near Nowra, New South Wales. This fig has a foliage circumference in excess of the Abbot Street Giant Fig mentioned above and, though not as tall, it does have one branch 1.88m in circumference.
unnamed small-leaved fig – (*Ficus eugenioides*), Chichester State Forest, New South Wales. This fig tree is 50m tall with a crown diameter of 40 metres.
Wollemi Pines – (*Wollemia nobilis*), Wollemi National Park, New South Wales. Located within 150km of Sydney, this recently discovered stand of tall pine trees is a species previously known only in the fossil record. Its ancestors have survived for millions of years, and so its discovery has captured the world's imagination, being likened to that of finding a small dinosaur still alive.

Cazneaux Tree

Red tingle, WA

CHAPTER 8
THE SKY

THE SKY

Weather and Climate

The major feature of Australian skies is the weather. The weather is the state of the atmosphere at any point in time as determined by temperature, atmospheric pressure, humidity, wind speed and direction, degree of cloudiness, and so on. Collectively, when taken together and averaged out over many years, these conditions are described as climate.

The weather is a major controlling factor on the nature and appearance of the Australian countryside. In the long term it is responsible for the wearing down of the earth's surface - the erosion, transportation and deposition of soil and rock particles, which contribute to producing the different landforms within the landscape. The major influences involved are wind, and water from precipitation – precipitation being any moisture that falls from the sky, in the form of rain, snow, sleet or hail.

The weather also exerts an influence on the plant and animal life. The variations in temperature, precipitation, humidity and wind all combine with the soil and rock types to produce formations of vegetation which characterise particular tracts of countryside at any point in time. Long-term changes in weather – in other words, climatic change – will result in long-term changes to these characteristic formations, which of course will have an impact on the fauna whose survival depends on specific plants or vegetation formations.

Day-to-Day Weather Changes

Day-to-day changes in the weather are most apparent during calm conditions when the weather is under the influence of high pressure systems – the most noticeable feature being differing temperatures between day and night. The heat received by the earth from the sun during the day is lost during the night, with consequent cooler temperatures. If there is a protective blanket of cloud to intercept this outgoing radiation, this heat loss at night is reduced. Conversely, cloud cover during the day will intercept incoming heat from the sun, producing a lower temperature than would otherwise be expected if the skies were clear.

Temperatures under these conditions are also affected by local winds, themselves formed in part by these day-to-day temperature differences. Temperatures are lowered by on-shore winds such as sea breezes, especially at times when oceanic waters are at their coolest. Conversely, temperatures may be raised by local winds blowing from the land or down the lee-side of mountain ranges. These conditions are most apparent during calm weather when the countryside is under the influences of high pressure systems. As the high pressure system moves away the daily weather falls under the influence of troughs and low pressure systems, which may bring cooler temperatures, rain or other precipitation, and winds.

Bondi Beach, NSW

THE SKY

Season-to-Season Weather Changes

Seasonal variations are more complex and trying to understand the reason as to why they occur involves looking at differences in air pressure (the highs and lows on the weather map) and the positions of these pressure systems over the country at different times of the year.

High and low pressure systems shift to the north during autumn and winter, and to the south during spring and summer. This shift follows, allowing for an appropriate time lag of about two months, the similar apparent movement of the sun. During the second half of the year the sun moves south and then, on the day of the summer solstice (20-22 December, depending on the number of years before or after the leap year adjustment), centres itself over the Tropic of Capricorn. Afterwards it appears to head north, arriving over the Tropic of Cancer on the day of Australia's (that is, the southern hemisphere's) winter solstice (20-22 June). Then the sun heads south again. The earth's atmospheric pressure systems follow the same pattern of north and south movement.

From around November to April, the 'summer half' of the year, high pressure systems move across Australia from west to east along the southern fringes of the continent. These high pressure systems direct easterly winds over much of the land. Usually fine weather is experienced at this time. As the high pressure system passes, the winds tend to become north-westerly, directing warm to hot air onto the southern regions. If the highs are slow moving or become stationary then the prevailing winds, again north-westerly, bring heat-wave conditions to southern Australia. Extreme maximums have been recorded at this time. In the far north of the country moist air moves in from the equatorial regions, bringing monsoonal (wet season) conditions.

Between May and October, the 'winter half' of the year, the high pressure systems still pass over the continent but take a more northerly path across the interior. Often they remain stationary for several days bringing superb sunny days followed by cool, even frosty nights. At this time northern Australia experiences dry and mild south-easterly winds (its dry season), while southern Australia is exposed to westerly winds and cold fronts associated with the low pressure systems that lie to the south of the high pressure belt. These fronts produce rainy and showery periods in southern Australia, though the south-east coast is often protected by the Great Dividing Range so that the days there tend to be cool but mostly sunny. If streams of cold air associated with these low pressure systems are directed on to the land, then cold, rainy or icy snow conditions will be experienced, especially in south-east Australia. These are the cold snaps of winter.

For a variety of reasons, which are only beginning to be understood, these seasonal variations do not behave the same way every year. During the winter of 1982 the high pressure systems – instead of centring themselves around 28°S latitude during late August, which is what might be expected – centred themselves around 34°S. This is where one would expect to find them during mid to late spring or mid autumn. As a result the weather experienced was more typical of spring or autumn than of winter. In eastern Australia that year the westerlies and cold fronts were limited to Tasmania and southernmost Victoria. This meant that a considerable part of New South Wales, northern and western Victoria and South Australia received little or no winter rainfall: drought, consequently, was declared over those parts of the country. This effect was repeated in 1994. It is now understood that El Nino – a term for the change in oceanic conditions in the Pacific Ocean – effects the pressure system patterns and, consequently, the weather experienced across the eastern half of Australia. An El Nino brings about below-average levels of rainfall. Its converse, La Nina, results in eastern Australia having higher than average rains, often for lengthy periods at a time. Significant La Nina conditions led to the flooding rains that inundated Charleville and Nyngan in 1990.

Australian weather is generally characterised by fine and sunny conditions over much of the country for most of the time. Nonetheless, despite the overall reasonably stable patterns of weather, variations do occur in the state of the skies and it is these that have the greatest impact on the Australian environment – producing events such as bushfires, floods, droughts and storms.

Table 8.1 Average Daily Sunshine in the Capital Cities

		over the year	in January	in June
NSW	Sydney	6.7hrs	7.2hrs	5.2hrs
Vic	Melbourne	5.7hrs	8.1hrs	3.4hrs
Qld	Brisbane	7.5hrs	7.5hrs	6.6hrs
SA	Adelaide	6.9hrs	9.9hrs	4.2hrs
WA	Perth	7.9hrs	10.1hrs	4.8hrs
Tas	Hobart	5.9hrs	7.9hrs	3.9hrs
NT	Darwin	8.5hrs	5.9hrs	9.7hrs
ACT	Canberra	7.2hrs	8.9hrs	4.8hrs

The area in Australia that has the least of sunshine is south-west Tasmania, with a yearly average of 4.8 hours per day. The sunniest areas, with a yearly average of 9.6 hours per day, are the north-west of Western Australia, and the eastern and northern parts of Central Australia, in the Northern Territory.

Table 8.2 Rainy Days and Clear Days in the Capital Cities

		rainy days	clear days
NSW	Sydney	148	85
Vic	Melbourne	143	48
Qld	Brisbane	123	98
SA	Adelaide	120	83
WA	Perth	120	108
Tas	Hobart	162	83
NT	Darwin	97	22
ACT	Canberra	110	121

Clear days are days virtually free of cloud. Note in table 8.2 that Sydney has more rainy days than Melbourne on average and that Hobart has the most number of rainy days of all the capitals. Perth is the wettest capital over the winter months.

The reason that Melbourne seems to be a rainier place than Sydney, even though it has barely more than half the average annual rainfall, is that much of Melbourne's rain falls as a steady drizzle over a period of days while Sydney's rain often falls as heavy but short showers. The rainier impression of a wet Melbourne is further enhanced by the relative consistency of its rainfall figures over a period of years (due in part to its southerly exposure to Bass Strait), while Sydney experiences a period of wetter years followed by a period of drier years (in part owing to the El Nino/La Nina effect).

Queensbury St, Melbourne

8.3 Locations Recording the Highest Yearly Average Temperature in each State/Territory

State/Territory		
NSW	Tibooburra	20.6°C
Vic	Mildura	16.9°C
Qld	Normanton	27.4°C
SA	Oodnadatta	21.8°C
WA	Wyndham	29.9°C
Tas	Eddystone Point	13.3°C
NT	Jabiru	28.0°C
ACT	Duntroon	13.4°C
OSA	Cocos Islands	26.9°C

The figures in table 8.3 are the average of maximum and minimum temperatures recorded throughout the year, over many years.

Marble Bar, Western Australia, is normally considered to be Australia's hottest town, but it can only manage an average temperature for the year of 27.6°C. Even its average maximum temperature for the year, 35.3°C, would still not make it Australia's hottest town (see comments beneath Table 8.4): those honours go to Wyndham and Camballin, which have average maximums of 36.2°C and 36.3°C respectively.

DID YOU KNOW?

Based on highest yearly average temperatures, **Australia's hottest** town is Wyndham. In 1946 Wyndham experienced maximum temperatures in excess of 32.2°C (the old 90°F) for 333 days in a row.

Wyndham, WA

Wyndham Port, WA

8.4 Hottest Location in each State/Territory during the Hottest Month

		hottest monthly average	average maximum temperature
NSW	Tibooburra	January average 29.1°C	35.9°C
Vic	Mildura	January average 24.3°C	31.8°C
Qld	Boulia	January average 31.6°C	38.5°C
SA	Oodnadatta	January average 30.3°C	37.8°C
WA	Nyang Station	February average 33.7°C	42.2°C
Tas	Launceston	January average 17.7°C	24.3°C
NT	Roper Bar	November average 32.1°C	39.2°C
ACT	Uriarra Camp	January average 23.5°C	31.6°C
OSA	Willis Island	January average 28.0°C	30.4°C

The figures in table 8.4 show the hottest monthly average (based on maximum and minimum temperatures) as well as the average maximum for that month. In other words, the locations represent the hottest places in each state/territory during the hottest month of summer.

DID YOU KNOW?

Marble Bar, Western Australia, is certainly **Australia's hottest town over the summer months**: in December, its hottest month, it records an average monthly temperature of 33.5°C, with a December average maximum average of 41.7°C – this is only beaten by Nyang Station listed above. On average, Marble Bar has 107 days per year with maximum temperatures in excess of 40°C. Between 31 October, 1923 and 7 April, 1924 there were 113 consecutive days of temperatures over 38.4°C (100°F) in Marble Bar – **a world record heatwave**.

Around Nyang Station is one of **the hottest areas in Australia** during the hottest months. There they can expect maximum temperatures in excess of 44.4°C (112°F) during February and 45.8°C (114.4°F) in March on at least one day in seven. In 1998 average summer maximum temperatures were 43.4°C at Nyang Station, culminating in an extreme maximum of 49.4°C on 16 February, the highest February maximum temperature recorded in Australia.

As a comparison, the coldest place in Australia at this time of the year is Mt Wellington, near Hobart, Tasmania, with an average temperature during January of 8.5°C and an average minimum of 4.3°C.

Sign, Marble Bar, WA

THE SKY

8.5 Locations Recording the Lowest Yearly Average Temperature in each

State/Territory		
NSW	Crackenback, near Thredbo	4.2°C
Vic	Mt Hotham	6.4°C
Qld	Applethorpe Research Station, near Stanthorpe	14.5°C
SA	Stirling	13.1°C
WA	Mettler, north-east of Albany	14.8°C
Tas	Mt Wellington	4.2°C
NT	Kulgera	20.6°C
ACT	Gudgenby Station	9.5°C
OSA	Macquarie Island	4.6°C

DID YOU KNOW?

Australia's coldest town (as opposed to alpine resorts) is Cabramurra, New South Wales, which has a yearly average temperature of 7.7°C. Mt Wellington, Tasmania, is **Australia's coldest place** (excluding Heard and Macquarie islands).

Cabramurra, NSW

The figures in table 8.5 are the average of maximum and minimum temperatures recorded throughout the year, over many years. The estimated yearly average temperatures on Heard Island, Southern Ocean, are lower than those in table 8.5. There, yearly average temperatures are estimated at about 1°C at sea level and about -6°C near the summit of Big Ben, the island's highest mountain.

The coldest town in Tasmania is Waratah, with a yearly average of 8.4°C. By comparison, the old and now totally abandoned mining town of Kiandra, in the Snowy Mountains of New South Wales, was once Australia's coldest town, with a yearly average of 6.0°C. The coldest true alpine village (as opposed to an alpine service area) is the winter-snowbound resort of Charlottes Pass, New South Wales, which has a yearly average of 4.9°C.

8.6 Coldest Location in each State/Territory during the Coldest Month

		oldest monthly average	average minimum temperature
NSW	Crackenback	August average -2.6°C	-5.1°C
Vic	Mt Hotham	July average -2.3°C	-3.6°C
Qld	Applethorpe Research Station	July average 7.4°C	1.2°C
SA	Stirling	July average 7.5°C	5.0°C
WA	Wandering	July average 9.7°C	4.2°C
Tas	Mt Wellington	July average 0.0°C	-1.9°C
NT	Curtin Springs	July average 11.2°C	3.2°C
ACT	Gudgenby Station	July average 3.0°C	-4.7°C
OSA	Macquarie Island	July average 3.0°C	1.4°C

The figures in table 8.6 show the coldest monthly average (based on maximum and minimum temperatures) as well as the average minimum for that month. In other words, the locations represent the coldest places in each state/territory during the coldest month of winter.

Heard Island, Southern Ocean, between June and October, has an average temperature of -2°C at sea level and an average minimum temperature of approximately -5°C.

As a comparison, the hottest settlement in Australia during the coldest month is Oenpelli, Arnhem Land, Northern Territory, which has a July average temperature of 25°C, and a July average maximum of 32°C.

Simpson Desrt, Qld

Tully, Qld

Table 8.7 Wettest Location of each State/Territory		
NSW	Tomewin, near Murwillumbah	2015mm
Vic	Wyelangta, Otway Range	1952mm
Qld	Bellenden Ker Top Station	8312mm
SA	Stirling, Mt Lofty Ranges	1189mm
WA	Sunnywest Farm, near Pemberton	1484mm
Tas	Lake Margaret, near Queenstown	3580mm
NT	Darwin Airport	1535mm
ACT	Jervis Bay	1170mm
OSA	Christmas Island	2518mm

The figures in table 8.7 represent annual average rainfall as measured in millimetres. It is possible that higher amounts fall on Mt Kosciuszko, New South Wales and in The Kimberley, Western Australia. It is possible that the new weather station at Mt Read, in western Tasmania, is wetter than Lake Margaret. In 1979 Bellenden Ker Top Station received 11,251mm of rain, approximately 443 inches in the old scale.

DID YOU KNOW?

In 2000 Bellenden Ker Top Station recorded 12,461mm – the **highest annual rainfall recorded in Australia**.

Australia's wettest town is Tully, in far north Queensland: it receives an average of 4321 millimetres per year spread over 154 rain days. Nearby Babinda is sometimes quoted as Australia's wettest town: its average is 4174mm per year over 151 rain days.

THE SKY

Table 8.8 Location with the Most Rainy Days in each State/Territory

NSW	Deer Vale, west of Dorrigo	157
Vic	Weeaproinah, Otway Range	206
Qld	Millaa Millaa	166
SA	Mt Gambier Airport	184
WA	Sunnywest Farm, near Pemberton	191
Tas	Waratah	253
NT	Darwin Airport	108
ACT	Canberra Airport	110
OSA	Macquarie Island	310

Rainy days are days when at least 0.1mm of rain falls. In Queensland the number of rainy days at Bellenden Ker Top Station probably exceeds the figure given in table 8.8. Nearby South Johnstone Experimental Station averages 178 rainy days per year while South Johnstone Post Office, 1.5km away, averages only 139 rainy days per year.

Table 8.9 Average Annual Rainfall in the Capital Cities

NSW	Sydney	1215mm
Vic	Melbourne	661mm
Qld	Brisbane	1157mm
SA	Adelaide	531mm
WA	Perth	879mm
Tas	Hobart	633mm
NT	Darwin	1536mm
ACT	Canberra	639mm

Note the relatively low figure for Hobart in table 8.9. This is the result of Hobart being in a rainshadow: the nearby bulk of Mt Wellington and the rugged mountains of south-west Tasmania intercept the prevailing moist westerly winds.

Table 8.10 Driest Locality of each State/Territory

NSW	Yandama Downs Station, near Tibooburra	150mm
Vic	Neds Corner, west of Mildura	246mm
Qld	Roseberth Station, near Birdsville	142mm
SA	Mulka Bore, Sturts Stony Desert	100mm
WA	Deakin Siding, Nullarbor Plain	174mm
Tas	Ross	457mm
NT	Charlotte Waters, Central Australia	128mm
ACT	Duntroon	534mm

At Forrest, on the Nullarbor Plain, Western Australia, the railway station receives 156mm per year, making it the driest locality; but the Meteorological Office, located only 1.5km away, receives an average of 182mm per year.

Oodnadatta, SA

DID YOU KNOW?

As with all statistics, nothing is straight forward. Oodnadatta, South Australia, is **Australia's driest town,** receiving an average of 117mm of rain per year. It has not been listed in table 8.10 because the figure for the nearby Oodnadatta Meteorological Office is 158mm per year.

It is rare that no rain falls in any one year at any one Australian locality. The only record found is of nil rain at Mylyie Station, Western Australia in 1924, but this may be unreliable. The **lowest rainfall on record** for any one year is 10.9mm recorded in 1889 at Mungeranie Bore, Sturts Stony Desert, South Australia.

8.11 Location with the Least Rainy Days in each State/Territory

NSW	Yandama Downs, near Tibooburra	13
Vic	Neds Corner, west of Mildura	39
Qld	Monkira Station, Channel Country	14
SA	Murnepowie Station, north-east of Lyndhurst	10
WA	Cunyu Outstation, near Wiluna	16
Tas	Chain of Lagoons, near St Marys	66
NT	Erldunda Station, Central Australia	15
ACT	Duntroon	81
OSA	Willis Island, Coral Sea	129

Desert Sand

Extreme Temperatures

Extreme temperatures are record-breaking temperatures that are well above or below expected maximums or minimums. Mostly, they are recorded during heatwaves, in the case of maximum temperatures, and cold snaps, in the case of minimum temperatures.

Table 8.12 Highest Recorded Temperature in each State/Territory

NSW	Wilcannia	50.0°C	10 January 1939
Vic	Mildura	47.2°C	10 January 1939
Qld	Birdsville	49.5°C	24 December 1972
SA	Oodnadatta	50.7°C	2 January 1960
WA	Mardie Station	50.5°C	19 February 1998
Tas	Hobart	40.8°C	4 January 1976
NT	Finke	48.3°C	1/2 January 1960
ACT	Canberra	42.2°C	11 January 1939

The figures in table 8.12 are the highest official maximum temperatures for each State/Territory. Higher figures have been recorded – see table 8.8 previous page. Even though the figure in table 8.8 were once considered to be the official extreme maximum temperatures there was some doubt about their reliability – this was due to the use of non-standardised instrumentation before 1910.

Many areas in western New South Wales, north-western Victoria, south-western Queensland and northern South Australia record high maximum temperatures during the summer months, partly as a result of the long trajectories of hot air emanating from Central Australia and northern Western Australia. For instance, near record state maximums were recorded during the February 2004 heatwave at Ouyen, Victoria – 46.7°C on 14 February; and Ivanhoe, New South Wales – 48.5°C on 15 February.

Wilcannia scene, NSW

THE SKY

Heatwaves

A heatwave is considered to be 5 or more consecutive days with temperatures at or above 35°C. During the summer of 2000-01 Marree, South Australia, recorded 47 days with maximum temperatures in excess of 40°C, culminating in an extreme maximum temperature for that summer of 47.2°C on 17 January.

A heatwave in 2004 saw near record extreme temperatures at Birdsville, Queensland, with a maximum of 48.5°C recorded on 5 January – the town's hottest day in 47 years. This heatwave saw Birdsville record an average maximum of 47.4°C over the first 6 days of January of that year.

A two week heatwave in northern Victoria, in February, 2004, saw a number of centres record up to 9 days with maximum temperatures at or over 40°C. This heatwave also covered parts of South Australia and New South Wales. The heatwave in New South Wales was considered to be one of the most severe ever experienced in that State, given its duration. Cobar, in western New South Wales, experienced 13 consecutive days (9-21 February) where the maximum temperature was over 40°C.

Table 8.13 Former Official Highest Recorded Temperature in each State/Territory

NSW	Bourke	52.8°C	17 January 1877
Vic	Mildura	50.8°C	6 January 1906
Qld	Cloncurry	53.1°C	16 January 1889
SA	Oodnadatta	50.7°C	2 January 1960
WA	Eucla	50.7°C	22 January 1906
Tas	Bushy Park	40.9°C	26 December 1945
NT	Charlotte Waters	48.2°C	2 January 1960
ACT	Canberra	42.2°C	11 January 1939
OSA	Cocos Islands	34.5°C	n.a.

DID YOU KNOW?
Worthy of note is White Cliffs, in western New South Wales. It has the **greatest temperature range** recorded in Australia: a maximum of 50.2°C and a minimum of -7°C, giving a range of 57.2°C.

The explorer Sturt, according to his diaries, recorded an unofficial temperature of 55.6°C near Strzelecki Creek, South Australia, in 1845. The temperature in the sun was 69.4°C. An unofficial temperature of 53°C has been recorded at Cocklebiddy, Nullarbor Plain, Western Australia.

Canberra, ACT

Table 8.14 Highest Recorded Temperatures in the Capital Cities

NSW	Sydney	45.3°C	14 January 1939
Vic	Melbourne	45.6°C	13 January 1939
Qld	Brisbane	43.2°C	26 January 1940
SA	Adelaide	46.1°C	12 January 1939
WA	Perth	46.2°C	23 February 1991
Tas	Hobart	40.8°C	4 January 1976
NT	Darwin	40.5°C	17 October 1892
ACT	Canberra	42.2°C	11 January 1939

Black Friday, Vic

Note in table 8.14 which capital has the lowest of the extreme maximum temperatures – the tropical city of Darwin. Tropical regions, particularly those on the coast, are typically very warm to hot all year but it is in the southern regions that higher extremes are obtained, owing to the long trajectory of warm air reaching across Australia.

The effect of the summer heatwave in 1939, resulting in the Black Friday bushfires in Victoria, can be clearly seen with three State capitals recording the highest-ever temperatures over three successive days.

During the summer of 2000-01 Adelaide experienced 29 days with maximum temperatures in excess of 35°C, its hottest summer in nearly 100 years. This was followed by another hot summer in 2003-04 when Adelide experienced 17 consecutive days of maximum temperatures at or over 30°C – a record – and an 8 day heatwave, from 13-20 February, with maximum temperatures at or over 35°C, including a 14 February maximum of 44.3°C, Adelaide's highest ever February temperature.

8.15 Highest Recorded Minimum Temperatures in each State/Territory

NSW	Ivanhoe	34.0°C	21 December 1994
Vic	Mildura	30.7°C	2 January 1955/7 January 1999
Qld	Birdsville	34.5°C	30 January 2003
SA	Arkaroola	35.5°C	24 January 1982
WA	Wittenoom	35.5°C	21 January 2003
Tas	Strahan	27.3°C	15 February 1982
NT	Finke	33.6°C	29 January 1973
ACT	Canberra	26.2°C	22 February 1960

During the heatwaves of early 2004, Ivanhoe, New South Wales, recorded a minimum temperature of 33.2°C on 15 February. Oodnadatta recorded an overnight low of 34.2°C on 4 January, after a day that reached 47.1°C.

Table 8.16 Highest Recorded Minimum Temperatures in the Capital Cities

NSW	Sydney	30.1°C	9 January 1983
Vic	Melbourne	30.6°C	1 February 1902
Qld	Brisbane	26.6°C	27 December 1952
SA	Adelaide	33.5°C	? January 1982
WA	Perth	29.3°C	1 January 1997
Tas	Hobart	22.8°C	27 January 1954
NT	Darwin	29.3°C	1 January 1968
ACT	Canberra	26.2°C	22 February 1960

In tables 8.15 and 8.16 the highest minimum temperatures are recorded between 9am of one day and 9am of the following day. Adelaide's hottest night occurred during an extended drought period when Adelaide had seven consecutive days of maximum temperatures in excess of 37.8°C (100°F).

In January 1982, near the end of a record-breaking drought, Melbourne experienced a nighttime minimum temperature of 31°C at 5am. At 6am the temperature was 32°C, the hottest ever recorded there at that time. A cool change moved through the city about an hour later, reducing the temperature by 10°C and thereby robbing that day of the record. In January 1997 Melbourne recorded its second warmest night on record: 30.3°C.

Cullen Bay, Darwin

Table 8.17 Lowest Recorded Temperature in each State/Territory

NSW	Charlottes Pass	-23.0°C	29 June 1994
Vic	Mt Hotham	-11.1°C	15 August 1968
Qld	Stanthorpe/Warwick	-10.6°C	23 June 1961/12 July 1965
SA	Yongala	-8.2°C	20 July 1976
WA	Booylgoo Station	-6.7°C	12 July 1969
Tas	Shannon, Butler Gorge, Tarraleah	-13.0°C	30 June 1983
NT	Alice Springs	-7.5°C	12 July 1976
ACT	Canberra	-10.0°C	11 July 1971
OSA	Heard Island	-11.0°C	n.a.

Mt Hotham, Vic

Table 8.18 Lowest Recorded Temperatures in the Capital Cities

NSW	Sydney	2.1°C	22 June 1932
Vic	Melbourne	-2.8°C	21 July 1969
Qld	Brisbane	2.3°C	12 July 1894/2 July 1896
SA	Adelaide	0.0°C	24 July 1908
WA	Perth	1.2°C	7 July 1916
Tas	Hobart	-2.8°C	25 June 1972
NT	Darwin	10.4°C	29 July 1942
ACT	Canberra	-10.0°C	11 July 1971

The figures in table 8.18 represent the lowest temperatures recorded at the main city weather stations; lower temperatures have been recorded in the suburbs. For instance: -4.6°C at Pennant Hills, Sydney; -5.0°C at Aspendale, Melbourne; -3.3°C at Sandgate, Brisbane; -3.0°C at Parafield, Adelaide; -1.6°C at Guildford, Perth; and -6.7°C at The Springs, Hobart.

8.19 Lowest Recorded Maximum Temperatures in each State/Territory

NSW	Crackenback	-6.9°C	9 July 1978
Vic	Mount Buller	-6.2°C	5 September 1995
Qld	Wallangarra	2.4°C	3 July 1984
SA	Mount Lofty	3.1°C	30 July 1994
WA	Narrogin	6.3°C	5 August 1973
Tas	Mt Wellington	-5.0°C	5 September 1995
NT	Yulara	5.9°C	11 July 1997

Alice Springs, NT

Table 8.20 Lowest Recorded Maximum Temperatures in the Capital Cities

NSW	Sydney	7.7°C	n.a.
Vic	Melbourne	4.4°C	n.a.
Qld	Brisbane	10.2°C	12 August 1954
SA	Adelaide	7.2°C	8 August 1936
WA	Perth	8.8°C	26 June 1956
Tas	Hobart	0.9°C	18 September 1951
NT	Darwin	18.6°C	14 June 1968
ACT	Canberra	2.8°C	28 June 1966

Hobart, Tas

Table 8.21 Some Cold Coastal Temperatures

Lowest coastal temperatures	
- on the mainland	-3.9°C at Eyre, WA
- in the tropics	-0.8°C at Mackay, Qld
- in Tasmania	-4.4°C at Swansea
- offshore	-11.0°C Heard Island, Southern Ocean

Swansea, Tas

THE SKY

Extreme Rain Events

Extreme rain events occur during storms, in the case of high rainfall, or droughts, in the case of low or nil rainfall.

Table 8.22 Highest Rainfall in One Day, in each State/Territory

NSW	Dorrigo	809mm	21 February 1954
Vic	Tanbryn	375mm	22 March 1983
Qld	Bellenden Ker Top Station	1140mm	4 January 1979
SA	Motpena Station, near Parachilna	273mm	14 March 1989
WA	Whim Creek	747mm	3 April 1898
Tas	Cullenswood, near St Marys	352mm	22 March 1974
NT	Roper Valley, Top End	545mm	15 April 1963

The fall at Whim Creek, cited in table 8.22, was due to a tropical cyclone; the average amount of rain there is normally 340mm per year.

On 17 August 1998 Wollongong, New South Wales, recorded 316mm in 24 hours, the vast majority of that falling over a 3 hour period. This resulted in devastating flash floods. In 2000, during Cyclone Steve, Bellenden Ker Top Station received 1050mm over 3 days, from February 25-28. In March 1996 Cape Tribulation, Queensland, received 765mm in 24 hours and 1230mm over a week. Redbank Mine, Northern Territory, received 1252mm of rain during the month of January, 2003.

DID YOU KNOW?

Bellenden Ker Top Station has also recorded the **highest monthly rainfall figure** of 5387mm, in January 1979.

8.23 Highest Rainfall in One Day, in the Capital Cities

NSW	Sydney	328mm	5 August 1986
Vic	Melbourne	108mm	29 January 1963
Qld	Brisbane	465mm	21 January 1887
SA	Adelaide	141mm	7 February 1925
WA	Perth	107mm	8 February 1992
Tas	Hobart	156mm	15 September 1957
NT	Darwin	296mm	7 January 1897
ACT	Canberra	126mm	15 March 1989

The Sydney figure, in table 8.23, resulted in the famous Sydney floods, which caused chaos as many roads were subjected to flash flooding. Sydney has also recorded 97mm in a one-hour period. In 1984, too, the Sydney suburb of Turramurra recorded 167mm in a two-hour period. There is an unofficial report of 518mm falling over 22.5 hours at South Head, Sydney, on 16 October, 1844.

8.24 Location Recording the Highest Annual Rainfall in each State/Territory

NSW	Tallowwood Point, North Coast	4540mm	1950
Vic	Falls Creek SEC, Victorian Alps	3738mm	1956
Qld	Bellenden Ker Top Station	12461mm	2000
SA	Aldgate, Mt Lofty Ranges	1852mm	1917
WA	Jarrahdale, Darling Range	2169mm	1917
Tas	Lake Margaret, near Queenstown	4504mm	1948
NT	Elizabeth Downs Station, Top End	2966mm	1973

The highest total falls of rain generally occur on mountain ranges that intersect prevailing moist air streams. This orographic uplift combined with a succession of particularly wet seasons has produced these damp extremes in table 8.24.

Falls Creek SEC, Victorian Alps

The Human Dimension - Impacts of Extreme Weather Conditions
Bushfires

Extended periods of hot dry weather or drought conditions dry out the land. The combination of dried-out vegetation with winds and low levels of humidity provides ideal conditions for bushfires. Bushfires are wild fires that burn in forests and shrublands; fires that burn in grasslands are called grass fires. A fire that is deliberately lit by authorities to burn up ground fuel (such as leaf litter, sticks, and so on) and maintained under some sort of control is a controlled burn, or hazard-reduction burn. Such burns are carried out to reduce the fierceness of dry season bushfires. In combating major conflagrations where property or persons are at risk, a fire deliberately lit by authorities is called a backburn.

The worst type of fire is a crown forest fire – when the crowns of trees are ablaze. Because the gasses released by the leaves are very volatile, the crown appears to explode. These exploding gases are propelled by the prevailing wind to such an extent that these fires, which can move at great speed (80km/hr or more), are virtually impossible to stop. Accompanying them,

Table 8.25 Some Significant Australian Bushfires

Year	Fire	Dates	Deaths	Buildings Destroyed	Area Burnt
1939	Black Friday	12-14 January	75	1100	2.5Mha
1961	Dwellingup	24 January	0	Dwellingup destroyed	?
1967	Hobart	7 Febuary	62	Over 3000	0.27Mha
1974	Western NSW	Dec 73-Jan 74	3	50	5Mha
1983	Ash Wednesday	15 February	75	2500	0.35Mha
1994	Sydney	1-8 January	4	185	?
2003	Canberra	18 January	4	400	?

Bushfires have a major impact on the environment, especially in the short term. In February 1967 the Hobart bushfires destroyed approximately 1450 structures as well as taking 62 lives. In 1974 a fire burnt out an area in western New South Wales of approximately 5 million hectares, roughly equal to 6 per cent of the state. Victoria's Ash Wednesday bushfires, in February 1983, destroyed 2000 houses, took 75 lives and burnt 335,000 hectares of countryside.

By far the worst fires were those experienced in Victoria and New South Wales in 1939. In January of that year a blocking high brought heat wave conditions for a fortnight with inland temperatures up to 47.7°C. On 8 January, 43 homes were destroyed and two people died at Dromana, Victoria. On 10 January 100 homes were destroyed and 19 people died, including 12 at the Rubicon timber mill. On 12 January some Victorian settlements were completely isolated by fires while Adelaide recorded its extreme maximum temperature of 47.6°C. On January 13th (Black Friday) Melbourne recorded 45.6°C and wind gusts up to 110km/hr were experienced elsewhere. As a result of these extreme conditions 1000 homes were burnt out, phone lines, bridges, railway lines and public utilities were destroyed, and 50 people died across the State. In addition, approximately 1.5 million hectares were burnt, representing 6.5 per cent of the State. On 14 January Sydney recorded its extreme maximum of 45.3°C; as well, 6 people died and 1 million hectares of countryside was burnt. The Royal Commissioner into the fires, Leonard E.B. Stretton, reported that 'those fires were lit by the hand of man'.

Ash Wednesaay, Vic

there are spot fires. A *spot fire* is one that runs ahead of a crown fire, having been ignited by air-borne embers originating from the main fire. An especially violent crown fire is known as a blow-up fire. Crown fires can devastate all in their path; they are the fires of the newspaper headlines and usually result in loss of life and significant property damage.

Bushfires are a natural phenomenon, being ignited by lightning, and some types of vegetation have adapted to the presence of fires as a part of their ecology. More commonly though, people start fires. These fires are deliberately, or accidentally, started. Arsonists light deliberate fires. Accidental fires arise from controlled burns or backburns getting out of control, or by landholders failing to totally extinguish a hazard-reduction burn. Low intensity hazard-reduction burns can result in fallen logs catching fire and smouldering for weeks. In southern Australia, especially during spring, hot north-westerly winds can cause these smouldering logs

Droughts

A drought results from of a lack of adequate rain or from an extended period of dry weather. During a drought streams stop flowing. Equally, the water transpired by plants or produced by evaporation from the soil is inadequate to sustain plants and animals: so, only the most drought-resistant plants can survive.

Droughts occur frequently in Australia and most native plant life is usually well-equipped to deal with them. Introduced crops and animals can be severely affected though, leading to crop failures, minimal planting of new crops, and the demise of introduced stock. Consequently, droughts have the most impact in areas given over to cropping or intensive grazing.

Drought-like conditions are common in areas that have extended dry seasons, such as the monsoonal tropics (no rain for 7-8 months), or Mediterranean-type climates (Adelaide and Perth, which are virtually rain-free for 3-4 months over summer). Droughts are least common in areas of regular winter rainfall – for instance, southernmost Victoria and Tasmania.

The droughts considered here are those ones in which the variability of rainfall is such that rainfall is equal to, or above average in some years, while in others it is so below the average that there is insufficient water for crops, stock or native vegetation. In economic terms, over the last 140 years there have been at least ten major droughts that have affected large parts of Australia – their severity usually being determined by the activity carried on within the drought area.

When the rains do come after extended drought conditions, they are often ineffective. This is because the ground surface, being unprotected and so dry, forms a hard crust: the rain simply runs off, often resulting in soil erosion. Only moderate soaking rains can effectively break a drought. Sometimes a light fall of rain or drizzle will fall over drought-affected crop or pasture lands, resulting in a greening of the landscape by small plants which are inadequate for pasture use. This is a so-called green drought.

Drought landscape, Bungonia, NSW

Table 8.26 Some Significant Australian Droughts

Years	Effects
1901-03	Australia's sheep flocks and cattle herds' numbers halved.
1911-16	Death of 19 million sheep and 2 million cattle.
1939-45	Death of 30 million sheep.
1958-68	Death of 20 million sheep; 40% decline in wheat harvest.
1982-83	Worst drought in eastern Australia.
1991-95	Severe drought in eastern Australia.
2002-03	Moderate to severe drought over much of Australia.

Floods

A flood is the inundation of land by water. It is caused by the inflow of tides or by a stream or streams, which have overflowed their banks. Stream floods are caused by a number of factors, the most important being a high intensity of rainfall falling into a catchment or drainage basin. High intensity rains may occur as a result of storms – thunderstorms or tropical cyclones or rain depressions (the disintegrated remains of tropical cyclones). Some of the low pressure systems that cross southern Australia during winter and early spring are also capable of producing high intensity falls, as are the east-coast lows that occasionally form off the New South Wales and southern Queensland coasts. In restricted high altitude localities, snow melting in spring can also produce floods. The likelihood of flooding increases in eastern Australia during La Nina years (opposite of El Nino), when there is a combination of an unstable atmosphere and prolonged rains.

The likelihood of floods is further enhanced when runoff conditions within the catchment have been modified, especially the wholesale clearance of native vegetation. The sudden and massive flooding of Charleville, Queensland, in 1990 would have been exacerbated by the widespread clearing of the Warrego and Nive river basins further upstream. Old townsfolk said they had never seen the river rise so fast. Other factors that act to exacerbate floods include infilling or draining of back swamps (which are natural flood mitigators), or when floodwaters meet high tides in estuaries, or storm surges (which are raised sea levels brought about by low atmospheric conditions, such as those found beneath tropical cyclones). Storm surge heights can raise sea levels up to 7m or more.

The amount of rain that has previously fallen and the degree of inundation that this has already caused are also critical factors. Owing to the behaviour of water flowing through catchments, when all flooded tributaries coalesce they produce one or more flood peaks. A flood peak can be likened to a long low wave of water moving downstream and its height will determine how widespread the flood is and what damage might result.

A flash flood is a special type of flood, common in mountainous terrain and hilly areas devoid of vegetation cover, such as in arid areas. It is also common in urban areas where many streams are channelled and most of the vegetation cover has been replaced with impervious concrete, bitumen and buildings. Flash floods are dangerous, for the water rises very rapidly and the flood peak passes quickly. In 1998 Wollongong, New South Wales, experienced torrential rains that resulted in flash floods. As the runoff tore down the Illawarra escarpment and entered the lower catchment creeks on the coastal plain, numerous buildings and roads were inundated, and considerable damaged caused.

The impact of floods can be beneficial, as it carries rich alluvium onto floodplains, which results in good pastures in low-lying country. But floods can also be devastating, when rapidly rising rivers pass through low-lying towns, or the country remains inundated for very long periods, stranding and drowning stock.

Table 8.27 Some Significant Australian Floods

Date	Flood
March, 1910	Todd River, Alice Springs, Northern Territory
April, 1929	north-eastern Tasmania floods
December, 1934	La Trobe River, Victoria
February, 1955	Hunter River, New South Wales
January, 1974	Brisbane River, Brisbane, Queensland
April, 1990	Warrego River, Charleville, Queensland
April, 1990	Bogan River, Nyngan, New South Wales
October, 1993	north-eastern Victoria floods
January, 1998	Katherine River, Katherine, Northern Territory

In February 1955, the Hunter Valley floods followed the dropping of 250mm of rain in 24 hours onto an already very wet region. Rapidly rising rivers (including the westward-flowing Macquarie, Gwydir and Namoi) flooded streets and houses, completely submerging houses in East Maitland – over 15,000 people were evacuated and 14 people died.

The Brisbane floods of 1974 followed probably the most rainfall to have fallen across Australia since European settlement. In Brisbane, minor flooding was exacerbated by tropical Cyclone Wanda dropping 580mm of rain over 3 days, with over 1200mm falling within the Brisbane River catchment. Homes bordering streams were washed away and 14 people died, some trapped by rising floodwaters.

Flooded valley, Bamarang, NSW

Storms

Storms occur when the stable condition of the atmosphere has been disturbed. Storms are manifested by very strong winds and are usually accompanied by rain or snow, thunder and lightning, and hail. The damage caused by storms can be total.

The most common storm is the thunderstorm. It occurs most frequently in northern Australia – with up to 60 thunder days per year in the Darwin region of the Northern Territory, compared to a maximum of 20 thunder days per year in southern Australia.

A thunderstorm is formed when heated air rises rapidly in the atmosphere and the water vapour condenses at high altitude into water and ice. This frozen water, unable to be held aloft, then falls and in the process creates a down-draft of air. As it nears the earth's surface, this down-draft spreads out horizontally, creating strong gusts called squalls. Each uplift and down-draft of air is called a cell, and a thunderstorm may contain many cells in a huge cloud called a cumulonimbus. Cells rarely last more than 20 minutes but new cells may replace old ones, so a storm can last some hours.

When many thunderstorms come together into a single, highly organised storm, this is called a super-thunderstorm. Such storms are much more destructive: they may last over 12 hours, producing widespread flooding rains, damaging hail and many squall lines.

Willy-willies are very destructive, tornado-like winds, and are related to thunderstorms. These winds are distinguished by funnel-shaped masses of air and uplifted material rotating very fast around a central axis. Fortunately, they are short-lived and their paths across the ground are very narrow: usually less than 500m wide. Consequently, in sparsely inhabited areas they may pass by unnoticed but in urban areas their destructive presence is very apparent. A willy-willy passing over water forms a waterspout. Willy-willies should not be confused with dust devils, which are commonly seen in dry country (even car parks) on hot days. These small, whirling vortexes of dust and other raised materials are relatively harmless.

On a much larger scale are the most devastating storms of all – the tropical cyclones. Particularly severe tropical cyclones, such a Cyclone Tracy, cause total destruction when passing over settled areas. These storms form and blow in the tropics; some however drift south into the temperate regions. Cyclones passing over land rapidly disintegrate and form rain depressions.

Tropical cyclones form during the 'summer half' of the year in tropical waters and are remarkable for their wind speeds. Once fully formed, a cyclone is distinguished by a huge swirling mass of cloud that reaches 10km in height and may spread nearly 300km kilometres from its centre. This cloudmass rotates in a clockwise direction (in the southern hemisphere) and at its centre is the eye: a clear or partly cloudy region, 20-60km across, with light winds and no rain. Within this cloudmass, and rotating around the eye, there are hurricane force winds – usually over 120km/hr, sometimes over 240km/hr – coupled with intense and widespread flooding rains.

Still larger than cyclones, although rarely as violent, are the storms of the temperate latitudes, the extratropical cyclones. Apart from those that form from southward migrating tropical cyclones, extratropical cyclones are associated with the cyclonic low pressure systems or depressions which cross southern Australia during the 'winter half' of the year. If sufficiently intense, these storms produce widespread gale-force winds – the dreaded 'winter westerlies' – that are accompanied by thunderstorms, lashing showery rains and, at altitude, sleet and snow. On the lee side of mountain ranges – typically the Great Dividing Range, which acts as a rainshadow – the winds are cold and dry. These windstorms can last for several days, with widespread damage resulting when the winds are sufficiently strong. In August, 2003, intense gales, with gusts up to 140km/hr, caused widespread damage and power blackouts in the southern and central New South Wales coastal areas.

The 'westerlies' – which are, in fact, mostly south-westerlies and sometimes southerlies – are often preceded by vigorous north-westerlies blowing off the inland. These winds are usually dry and can reach gale force at times. Passing over drought-affected land, bare paddocks or arid areas, they raise dust and in the process produce dust storms. The immediate effects of dust storms include reduced visibility, a choking atmosphere and the deposition of dust. In the main, dust storms are a consequence of non-Aboriginal occupation – of land clearances and overgrazing.

As a south-westerly approaches there is a sudden change in wind direction, a drop in temperature, often squalls and occasionally thunderstorms. These sudden changes are called cold fronts or, if less intense, cool changes. These fronts and their preceding winds may also cause widespread damage, especially during late winter-early spring, and also raise dust, forming dust storms. Cool changes also herald the cool southerly winds of summer, some of which may be vigorous and storm-like. But not all changes are storm-like: in the stable atmosphere of late summer-autumn some changes are so weak they might pass through without notice.

THE SKY

DID YOU KNOW?

The strongest wind speeds recorded include: 259kms/hr at Mardie Station, Western Australia during Cyclone Trixie; 267km/hr at Learmonth, Western Australia, during Cyclone Vance – this is the **strongest wind speed** recorded in Australia.

Table 8.28 Wind Speeds and Strongest Recorded Wind Gusts in the Capital Cities

		Average Wind Speed	Strongest Gust
NSW	Sydney	11.6kms/hr	153kms/hr
Vic	Melbourne	12.3kms/hr	119kms/hr
Qld	Brisbane	10.8kms/hr	128kms/hr
SA	Adelaide	12.4kms/hr	148kms/hr
WA	Perth	15.6kms/hr	156kms/hr
Tas	Hobart	11.5kms/hr	150kms/hr
NT	Darwin	9.2kms/hr	217kms/hr
ACT	Canberra	5.8kms/hr	128kms/hr

Richmond, New South Wales, on the outskirts of Sydney, recorded a wind gust of 174km/hr during a severe thunderstorm on 3 December, 2001 – the highest wind gust recorded in that State. The figure in table 8.28 for Darwin was recorded during Cyclone Tracy.

References have been made to some unofficial wind speeds. In December 1998 Cyclone Thelma was probably Australia's strongest cyclone of the 20th century. When passing over Melville and Bathurst islands, Northern Territory, wind gusts were estimated to be up to 295km/hr. A cyclone passing over Pardoo Roadhouse, Western Australia, in February 2002, also experienced estimated wind gusts up to 295km/hr. Cyclone Rosita, in April 2000, reached estimated wind gusts up to 290km/hr as it destroyed a tourist resort at Eco Beach, south of Broome, Western Australia. Cyclone John, in December 1999, passed over the Pilbara coast destroying the township of Whim Creek, Western Australia, with no injuries. Wind speeds were estimated at over 300km/hr.

In January 1991 a tornado swept through the upper North Shore suburbs in Sydney, New South Wales, during which wind speeds are estimated to have reached 290km/hr. On 15 December 1967 a tornado cut a swathe of destruction through the Sydney suburb of Mosman. In some 90 seconds a funnel some 600m high and 45m wide extensively destroyed 250 buildings; wind speeds were estimated at 200km/hr.

Table 8.29 Some Significant Australian Storms

Year	Storm	Effects
1908	Broome Cyclone	devastating winds/flooding rains/storm surge, 50 killed
1918	Mackay Cyclone	devastating winds/flooding rains/storm surge, 30 killed
1967	Mosman Tornado	devastating winds/250 buildings destroyed
1970	Cyclone Ada	devastating winds/Whitsunday holiday resorts destroyed, 13 killed
1974	Cyclone Tracy	devastating winds/Darwin mostly destroyed, 65 killed
1999	Sydney Hailstorm	squally winds/tennis ball sized hailstones/thousands of roofs and cars damaged, no deaths

Cyclone Tracey devastation, Darwin (top and bottom)

Table 8.30 Incidence of Snow: by State/Territory

New South Wales
Regular winter ground cover above 1400m along the Great Dividing Range; occasional falls on the lower hillslopes, particularly above 300m.

Victoria
Regular winter ground cover above 1400m along the Great Dividing Range; occasional falls on the lower hillslopes; rare falls over many of the southern Victorian hills.

Queensland
Occasional falls along the Great Dividing Range as far north as the Bunya Mountains; has been recorded at Roma; unofficial recordings in the Central Highlands. One reference mentions a fall of snow at Mackay, Queensland.

South Australia
Occasional falls on the highest peaks of the Mt Lofty and Flinders Ranges and in the Jamestown-Peterborough district.

Western Australia
Occasional falls on the summit of Bluff Knoll, Stirling Ranges; rare falls along the Darling Range; snow recorded as far north as Wongan Hills.

Tasmania
Regular winter ground cover above 1000m; regular and occasional falls at lower altitudes.

Northern Territory
Snow reported at Uluru, Central Australia.

Australian Capital Territory
Regular winter ground cover above 1400m; occasional falls on the lower hillslopes.

Off-Shore Australia
Regular snowfalls on Macquarie Island; Heard Island is mostly ice-clad.

The highest peaks of the Snowy Mountains, Victorian Alps and Tasmanian Highlands may receive snow at any time of the year. Regular snow patches may remain all year in the Mt Kosciuszko area, and have also be known to remain all year on the southern slopes of Mt Anne in south-west Tasmania. It used to be claimed that in winter Australia had more snow than Switzerland.

Snow will fall in air temperatures up to 7°C but will usually melt at once; if the temperature is lower than 3°C it will lie on the ground. Where the average minimum temperature of the coldest month is lower than -3°C it will remain on the ground for extended periods. At temperatures between -3°C and 4°C the ground cover may be patchy.

The temperature of crusted snow – that which has been subjected to freezing – can be determined by the noise it emits when walked upon: at around 0°C it makes a deep crunch; at -5°C it emits a high squeak; at -15°C, a very high squeak.

DID YOU KNOW?
The **greatest fall of snow** recorded over a seven day period was 1170mm at Spencers Creek, Snowy Mountains, New South Wales, in May 1964.

DID YOU KNOW?
Accumulations of snow can produce avalanches – especially hard-packed snow and ice, such as found in cornices located on the highest of the snow-covered peaks. One of the **biggest recorded avalanches** occurred in the winter of 1981 on the upper slopes of Mt Townshend, Snowy Mountains, New South Wales. It swept into Lady Northcotes Canyon and into the treeline below, flattening many trees and temporarily damming the valley. The path of destruction was still visible two years later.

Fallen snow, Mt Macedon, Vic

Sydney	Recorded on 14 days out of 146 years in parts of the metropolitan area.
Melbourne	Occasional falls on hills within 35 kilometres of the city.
Brisbane	Unlikely
Adelaide	Recorded on 120 days in 128 years at Mt Lofty summit.
Perth	Unlikely
Hobart	Regular snowfalls on the slopes of Mt Wellington during winter; occasional falls in the city, especially in the more elevated suburbs.
Darwin	Never
Canberra	Regular falls on the hills outside the city; occasional falls within the city.

The Sydney figures probably include soft hail that can be mistaken for snow. Snow has been sighted from the tops of skyscrapers in Sydney and Melbourne.

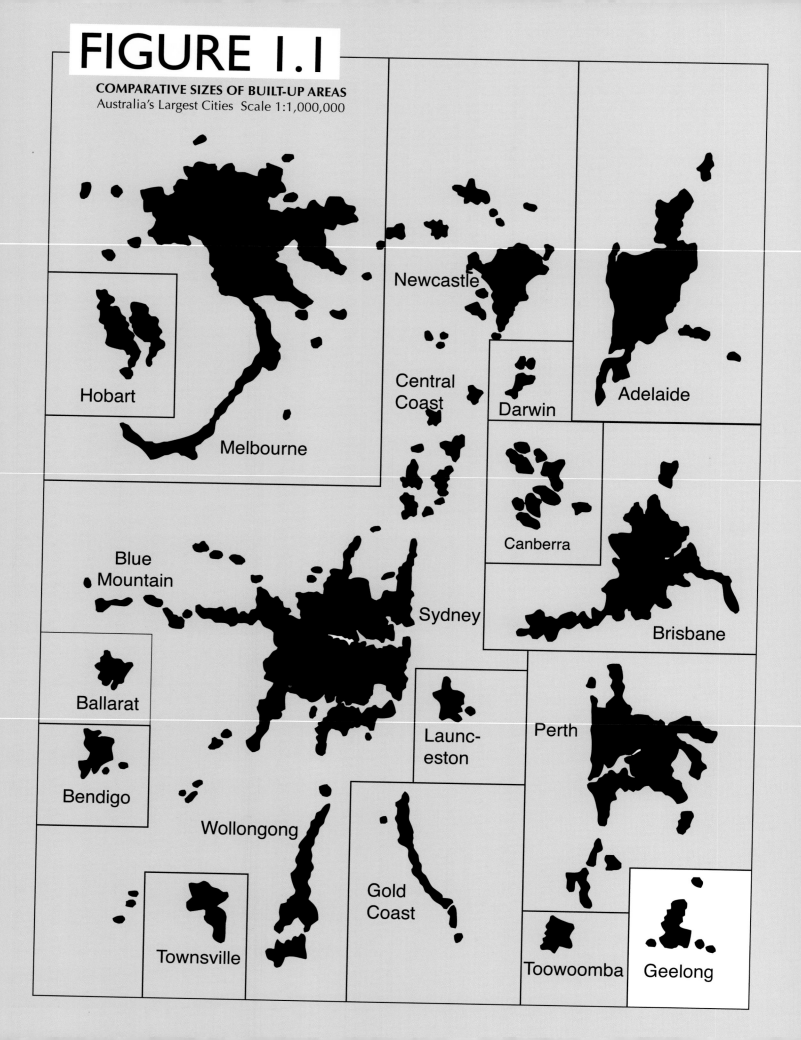

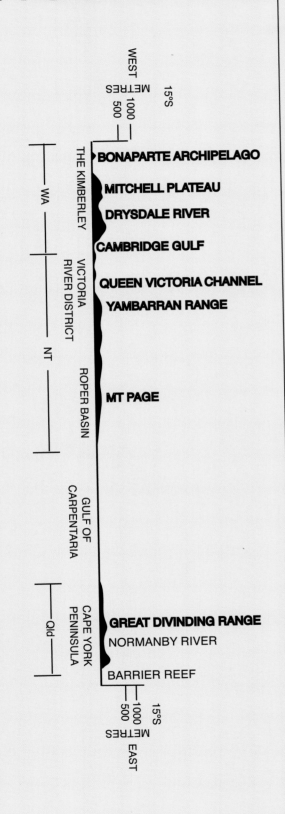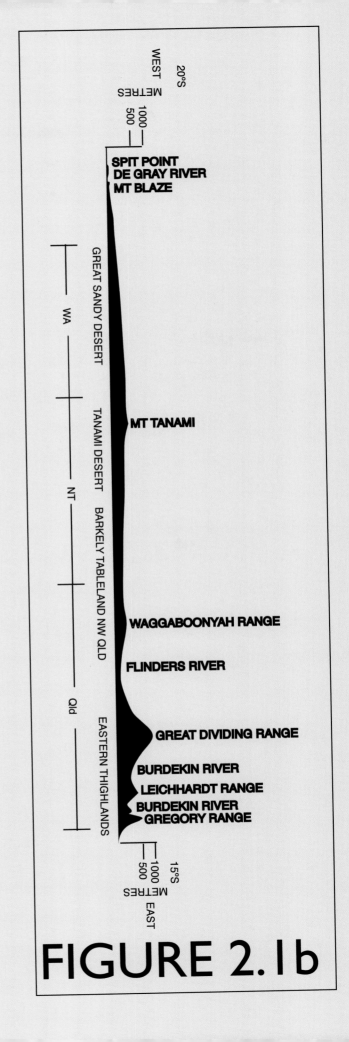

FIGURE 2.1c

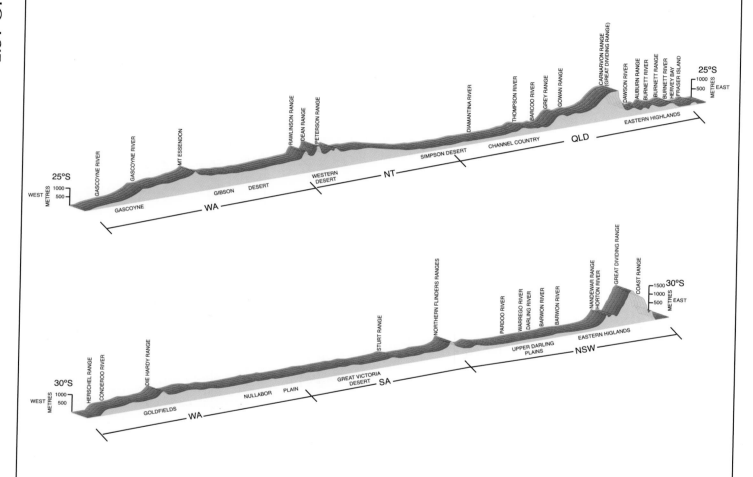

FIGURE 2.1d

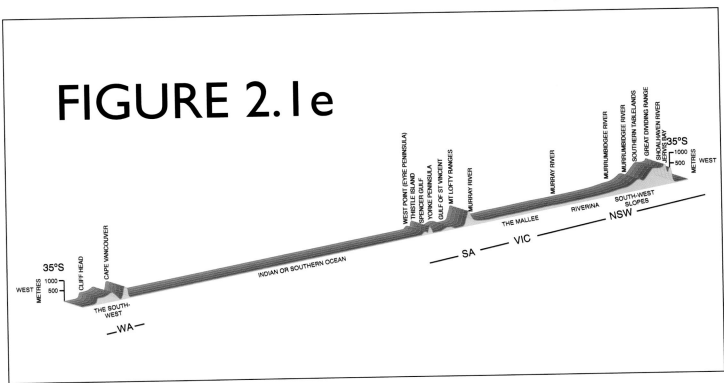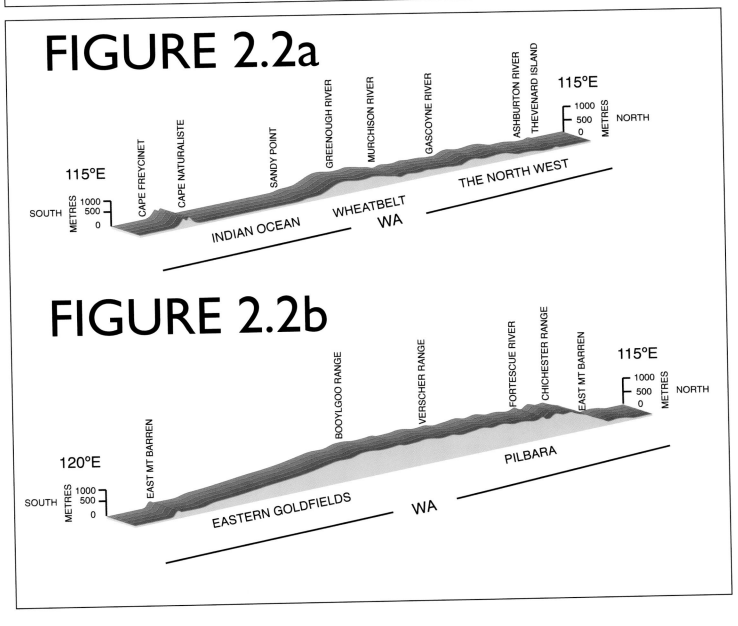

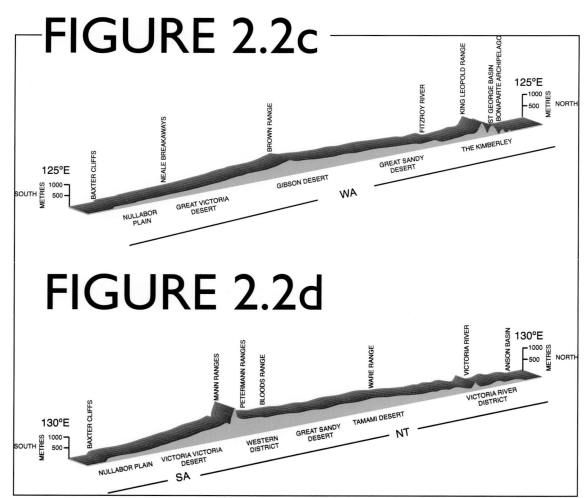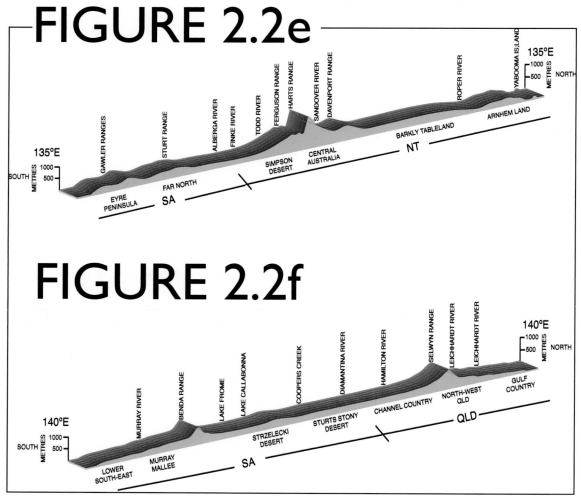

LIST OF FIGURES

FIGURE 2.2g

FIGURE 2.2h

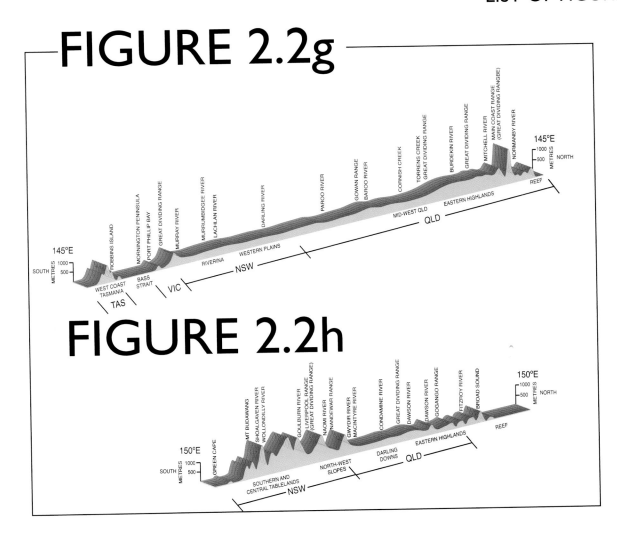

FIGURE 3.1

COMPARITIVE SIZE OF THE HIGHEST MOUNTAINS IN EACH STATE/TERRITORY

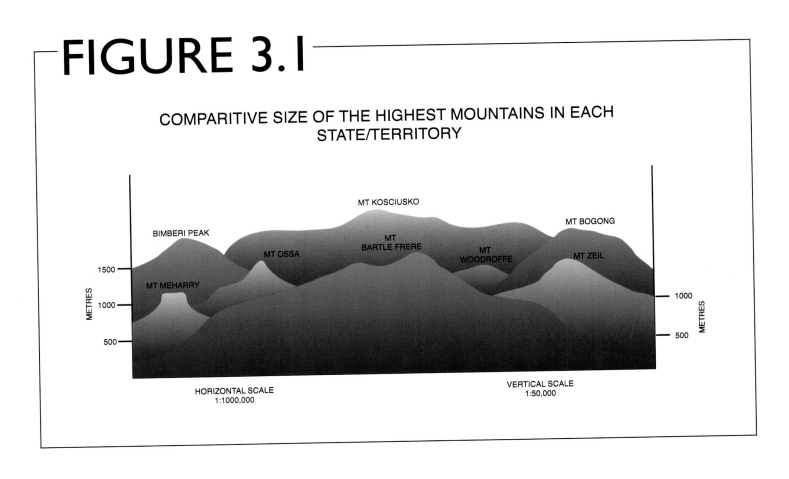

HORIZONTAL SCALE
1:1,000,000

VERTICAL SCALE
1:50,000

FIGURE 4.1

CROSS SECTIONS OF VALLEYS AND GORGES IN AUSTRALIA
Horizontal Scale: 1 to 100,000 Vertical Scale: 1 to 50,000 Base line indicate sea level.

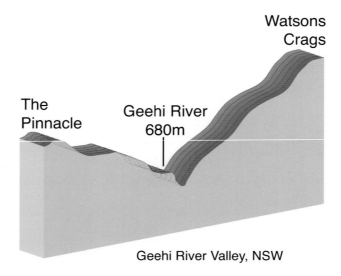

Geehi River Valley, NSW

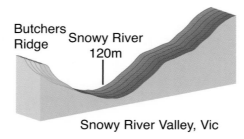

Snowy River Valley, Vic

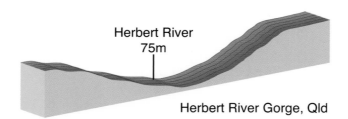

Herbert River Gorge, Qld

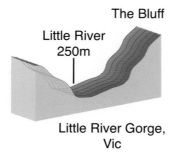

Little River Gorge, Vic

FIGURE 4.2

CROSS SECTIONS OF THE GREATEST HILLSLOPE IN EACH STATE/TERRITORY
Horizontal Scale: 1 to 100,000   Vertical Scale: 1 to 50,000   Base line indicate sea level.

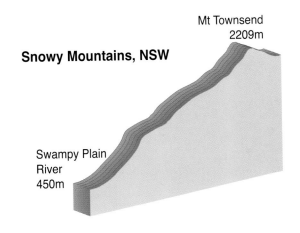

Snowy Mountains, NSW — Mt Townsend 2209m; Swampy Plain River 450m

Stirling Range, WA — Bluff Knoll 1096m; Dry Lake 140m

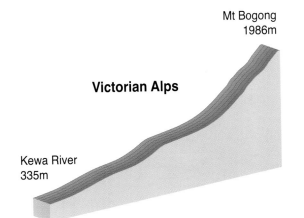

Victorian Alps — Mt Bogong 1986m; Kewa River 335m

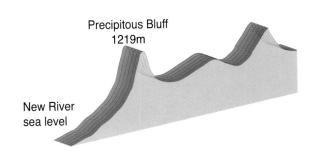

Southern Coast of Tasmania — Precipitous Bluff 1219m; New River sea level

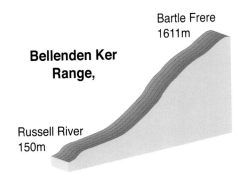

Bellenden Ker Range — Bartle Frere 1611m; Russell River 150m

Northern Territory - reliable information unavailable for accurate cross section

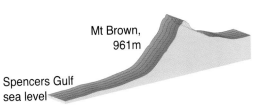

South Flinders Range, SA — Mt Brown, 961m; Spencers Gulf sea level

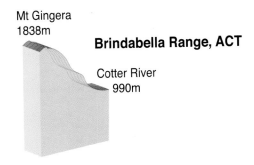

Brindabella Range, ACT — Mt Gingera 1838m; Cotter River 990m

FIGURE 4.3

CROSS SECTIONS OF RANGES AND VALLEYS NEAR THE CAPITAL CITIES
Horizontal Scale: 1 to 100,000 Vertical Scale: 1 to 50,000 Base line indicate sea level.

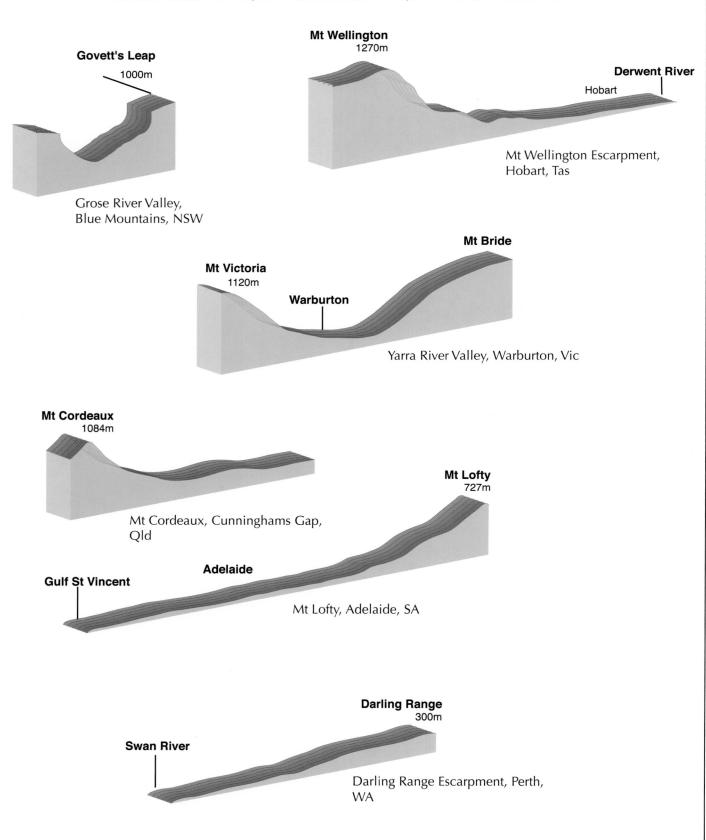

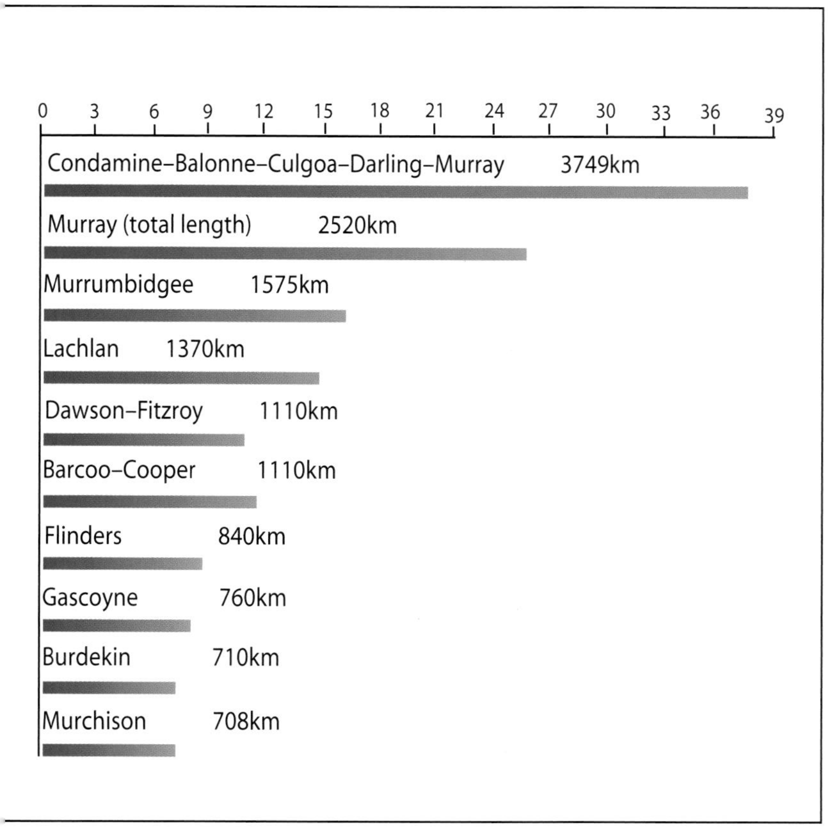
0 3 6 9 12 15 18 21 24 27 30 33 36 39
Condamine–Balonne–Culgoa–Darling–Murray 3749km
Murray (total length) 2520km
Murrumbidgee 1575km
Lachlan 1370km
Dawson–Fitzroy 1110km
Barcoo–Cooper 1110km
Flinders 840km
Gascoyne 760km
Burdekin 710km
Murchison 708km

FIGURE 5.2

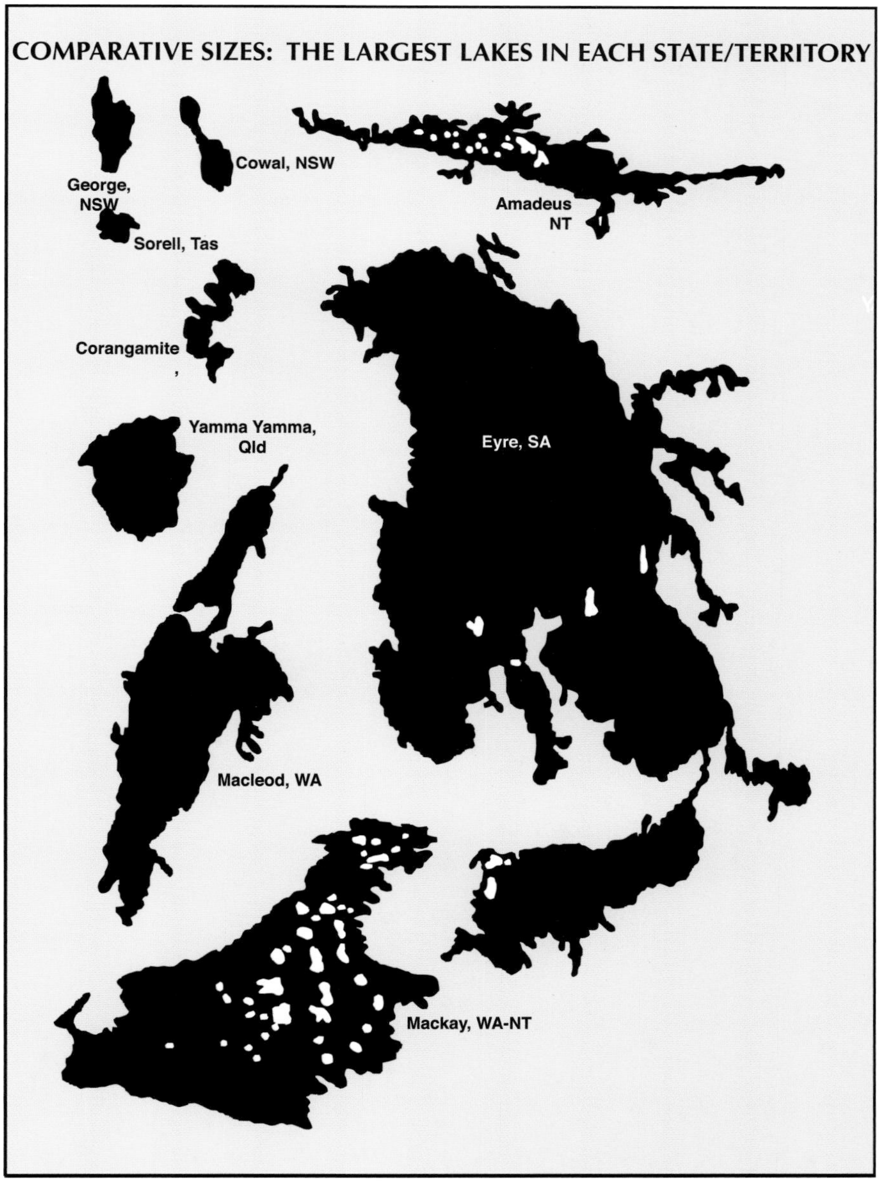
COMPARATIVE SIZES: THE LARGEST LAKES IN EACH STATE/TERRITORY
George, NSW
Cowal, NSW
Amadeus NT
Sorell, Tas
Corangamite
Yamma Yamma, Qld
Eyre, SA
Macleod, WA
Mackay, WA-NT

FIGURE 5.3

APPENDIX
AUSTRALIAN PLACE NAMES

Below are a number of place names regarded as interesting, unusual or not particularly well known. No doubt the earliest settlers named most of these places and it is interesting to contemplate how these names might have arisen. While some are self-explanatory others are somewhat less obvious.

Unusual Names
Blackfellow Bones Bore: bore north-east of Alice Springs, Northern Territory.
Boy in a Boat: seasonal swamp near Laverton, Western Australia.
Burrumbuttock: township near Albury, New South Wales.
Cadibarrawirracanna: salt lake east of Coober Pedy, South Australia.
Cardivillawarracurracurrieapparlandoo: old name of a bore in South Australia meaning 'reflection of the stars in the water'.
Come-by-Chance: small rural township near Walgett, New South Wales.
Crookwell: country town, Southern Tablelands, New South Wales.
Dead Bull Creek, Dead Calf Creek, Wild Cow Creek: succession of creek names on the Bonang Highway, Victoria.
Dead Dog Flat: locality near Mt Garnet, Queensland.
Emohruo: homestead in western New South Wales - try saying it backwards.
Howlong: country town on the Murray River, New South Wales.
Lake Muck: station near the South Australia border in western New South Wales.
Little Mother of Ducks: small lake near Guyra, New South Wales.
Mamungkukumpurangkuntjunya: hill in northern South Australia, and *Australia's longest officially named place.*
Mummarraboogungoorangil: large swamp near Duaringa, Fitzroy Basin, Queensland.
No Where Else: old locality name near Sheffield, Tasmania.
Seldom Seen: mountain name and small roadhouse at Wulgulmerang, Victoria.
Turn Back Jimmy Creek: creek in the Riverina, New South Wales.
Uardry: siding on the old Hay railway line, New South Wales.
Ulladulla: town on the south coast of New South Wales.
Useless Loop: salt mining settlement near Denham, Western Australia.
Wait-a-While: siding on the old Tocumwal railway line, New South Wales.
Wilsons Downfall: settlement near Tenterfield, New South Wales.
Wing Ding: station in western New South Wales.
Youldoo: station in western New South Wales.
Youltoo: station near Youldoo.

Towns and Localities with a Person's First Name
Albert, Victoria
Alexandria, Victoria
Barry, New South Wales
Donald, Victoria
Foster, Victoria
Keith, South Australia
Kelly, South Australia
Lawrence, New South Wales
Maude, New South Wales
Morgan, South Australia
Shannon, Tasmania
Tara, Queensland

Towns and Localities with the Name of an Object or Concept
Bell, New South Wales/Queensland
Berry, New South Wales
Bright, Victoria
Bluff, Queensland
Cargo, New South Wales
Casino, New South Wales
Comet, Queensland

APPENDIX

Crows Nest, New South Wales/Queensland
Cue, Western Australia
Cygnet, Tasmania
Derby, Western Australia/Tasmania
Dingo, Queensland
Hay, New South Wales
Prairie, Queensland
Rainbow, Victoria
Sale, Victoria
Sedan, South Australia
Speed, Victoria
Wombat, New South Wales

Australia's most common name, according to the 1:250,000 Map Gazetteer is Spring Creek with 347 mentions. Four Mile Creek (112 mentions), Six Mile Creek (100), Five Mile Creek (78) and Three Mile Creek (51) are the next most common place names.

REFERENCES

Australian Surveying and Land Information Group, 3rd series, 1980-1990, *Atlas of Australian Resources, Volumes One to Six*, AGPS, Canberra
Barrett, C, 1944, *Australian Caves, Cliffs and Waterfalls*, Georgian House, Melbourne
Bureau of Meteorology, 1977, *Manual of Meteorology*, AGPS, Canberra
Clancy, R, 1995, *The Mapping of Terra Australis*, Universal Press, Sydney
Commonwealth Year Books, various editions, AGPS, Canberra
Cranby, S, 1993, *Oxford Australia Student Atlas*, OUP, Melbourne
Groves, RL(ed), 1981, *The Vegetation of Australia*, Cambridge University Press
Elkin, AP, 2nd ed., 1980, *Aboriginal Men of High Degree*, UQP, Brisbane
Fitzgerald, 1986, *Java La Grande: The Portuguese Discovery of Australia*, The Publishers Pty Ltd, Hobart
Flood, J, 1995, *Archaeology of the Dreamtime*, Angus and Robertson, Sydney
Graetz, D, Fisher, R, Wilson, M, 1992, *Looking Back: The Changing Face of the Australian Continent, 1972-1992*, CSIRO, Canberra
Halliday, I & Hill, R, 1974, *A Field Guide to Australian Trees*, Rigby, Adelaide
Jeans, DN (ed), 1986, *Australia: A Geography, Volume One: The Natural Environment*, SUP, Sydney
Jeans, DN (ed), 1987, *Australia: A Geography, Volume Two: Space and Society*, SUP, Sydney
Johnson, K, 1992, *The Ausmap Atlas of Australia*, CUP, Melbourne
Kirkpatrick, J, 1994, *A Continent Transformed: Human Impact on the Natural Vegetation of Australia*, OUP, Melbourne
Laseron, CF, (revised by J.N.Jennings), 1972, *The Face of Australia*, Angus and Robertson, Sydney
Latz, P, 1995, *Bushfires and Bushtucker: Aboriginal Plant Use in Central Australia*, IAD Press, Alice Springs
Lawrence, DH, 1964, *Kangaroo*, William Heinemann Ltd, London
Learmouth, N & A, 1971, *Regional Landscapes of Australia*, Angus and Robertson, Sydney
Leeper, GW (ed), 1970, *The Australian Environment*, CSIRO, Sydney
Peasley, WJ, 1983, *The Last of the Nomads*, Fremantle Arts Centre Press, Fremantle
Ralph, B, 1996, *Longman Australian Atlas for Secondary Schools*, Addison, Wesley, Longman Australia, Melbourne
Read, IG, 1994, *The Bush: A Guide to the Vegetated Landscapes of Australia*, UNSW Press, Sydney
Readers Digest, 1994, *Readers Digest Atlas of Australia*, Readers Digest (Australia), Sydney
Recher, HF (ed), 1976, *Scenic Wonders of Australia*, Readers Digest Services, Sydney
Rodwell, P (ed), 1991, *Discover Australia*, Readers Digest, Sydney
State Year Books, various editions
Sturman, A, & Tapper, N, 1996, *The Weather and Climate of Australia and New Zealand*, OUP, Melbourne

All photographs by the author except: pages 7 (bottom), 9, 14, 19, 22 (top), 24 (right), 28-29, 30 (all), 32, 33 (all except Albany), 34, 35, 36 (all), 37 (top), 42 (top), 52 (top), 53 (all), 71, 72, 73 (all), 76, 82, 87, 88 (top), 90, 93 (all), 97 (top), 98, 102, 103 (top), 104 (left), 105, 106, 111, 113 (all), 116-117 (all), 118, 121 (top), 123 (all), 124 (top), 125, 126 (all), 127 (all), 128 (all), 131 (right), 132 (right), 134 (all), and 135 (all).

ACKNOWLEDGMENTS

We are very grateful to the following for their kind assitance in providing photographs for this publication:

National Library of Australia: pages 5, 7, 11–23, 106, 113.
Tourism NSW: pages 24, 32, 33, 48, 50, 52, 53, 55, 63, 70, 72, 79, 86, 87, 90, 105, 116, 121, 124.
Visions of Victoria: pages 48, 50, 53, 62, 64-67, 80, 88, 93, 127, 129, 132, 136.
South Australian Media Gallery: pages 33, 36, 37, 39, 40, 41, 46, 50, 54, 71, 77, 83, 84, 92, 93, 98, 123.
Commonwealth of Australia: page 33.
Department of Sustainability and Environment: pages 126, 131, 135.
Broken Hill Regional Tourist Association: page 36.
Fraser Coast South Burnett Regional Tourism Board: page 93.
Geoscience Australia: page 60.
Dale Simmons – Powerworks Energy Technology Centre: page 28.
Rob Hunt – NQ Water: page 88.
Kieran Wartle, Dirk Hartog Island: pages 30, 98
www.ozthunder.com: pages 129, 134, 135.

INDEX

151

101 Collins Street 24
120 Collins Street 24
4Qs Radio Tower 25

A
Abbot Street Giant Fig 112, 113
Abbott Peak 48
Aberfeldy 54
Aberfeldy-Walhalla Road 66
Aboriginal Occupation 12
Aborigines 13, 14, 15, 84
Abrakurrie Cave 72
Abt Railway 68
Acacia 103, 107
Ada Tree 106
Adelaide 18, 34, 60, 118, 123, 126, 127, 128, 129, 135, 137
Adelaide Plains 39
Adelaide River Bridge 23
Adelaide River Floodplain 109
Admiralty Gulf 91
Aileron 35
Albany 18, 33, 94
Alberga Dunefield 42
Aldgate 129
Alerumba Creek 62
Algebuckina 81
Algebuckina Bridge 81
Alice Springs 35, 54, 58, 113, 127
All Saints Church 21
Alligator Gorge Road 56
Allocasuarina 103
alluvial fans 76
alluvial plains 38
Alpine ash 105
Alpine herbfields 108
Alpine Road 56, 66, 67
Alpine Way 56, 67
Alpurrurulam 33
Altamount 24
Amadeus Basin-Darwin pipeline 42
Amata 54
Amata-Pipalyatjara Road 57
Amiens Road 56
AMP Centrepoint Tower 24, 25

Ancient Empire 112
Andamooka 85
Andromeda Stand 106
Anna Creek Station 40
Anne Beadell Highway 43
Anne-A-Kanada 72
Anser Group 30
Antarctic beech 104
Anzac Bridge 70, 79, 80
Apollo Bay 32
Applethorpe Research Station 121
Apsley Falls 83
Apsley Gorge 63
Apsley Strait 98
Arafura Swamp 109
Archaringa River 77
Areyonga 35, 54
Arkaroola 88, 126
Arltunga Tourist Drive 57
Arnhem Land 67
Arnhem Land Plateau 47
Arrernte 13
Arunta Desert 43
Ash Wednesday 131
Ashburton River 78
Ashmore Reef 17
Aspendale 127
Asteraceae 103
Atherton Tableland 47, 59, 86
Atila 51
Atriplex 103
Aurari Bay Beach 93
Australia's Plains 38
Australian Alps Walking Track 37
Australian Capital Territory 41, 47, 69
Australian Labor Party 113
Australian Landmass 28
Australian Portland Cement Company 69
Australian vegetation types 104
Ayers Rock 51, 52, 83
Aylmerton Tunnel 69

B
B4 - Buchan Show Caves 72
Babinda 122
Bacchus Marsh-Ingliston 68
Back Creek Viaduct 70
Backstairs Passage 98
Bajo fishermen 14, 15, 17
Bald Hills Transmitter Tower 25
Bald Rock 47

Ballarat 24
Ballera-Mt Isa pipeline 42
Ballera-Wallumbilla pipeline 42
Balls Pyramid 49, 59, 95
Balonne River 77, 88
Banjelang 13
Bank West Tower 24
Banks Strait 98
Banksia 103
Baobab 104, 105
Barcaldine 113
Barcoo River 77
Barked-bloodwood 106
Barkinji 13
Barkly Highway 37, 41
Barkly Tableland 38, 39, 73, 85
Barrabarrac Hole 73
Barrenjoey 93
Barrier Highway 37, 58
Barrington Plateau 47
Barrington Tops 59
Barron Falls 83
Barron Gorge 62
Bartle Frere South Peak 49
Barwon River 38, 77
Bass 14
Bass Strait 98, 118
Batavia 17, 20
Bathurst 18
Bathurst Island 98
Bathurst Islands 135
Batman Bridge 70, 80
Baudin 14
Baxter Cliffs 66, 67
bays 91
Beach dune formations 111
Beachport Jetty 96
Beadells Tree 112
Beardmore Dam 88
Beardy-Dumaresq River junction 30
Bedourie 85
Belalie North 58
Bell 54
Bellenden Ker South Peak 49
Bellenden Ker Top Station 122, 123, 129
Bellenden Ker Tower 52
Ben Lomond 53, 54, 58, 66, 67
Ben Lomond Plateau 47, 86
Ben Lomond Summit Road 57
Benambra 54
Bendoc 54

Berri Bridge 80
Bethangra Bridge 80
Bethungra 68
Beulah Homestead 55
Bherwerre Beach 93
Bibbulmun Track 37
Bicentennial National Trail 37
Bicheno 33
Big Ben 47, 49, 59, 121
Big Desert 43, 107
Big Fella Gum Tree 106
Big Hill 69
billabongs 76
Bilpa Morea Claypan 85
Bilpin-Mt Irvine Road 66
Bilson-Blue Cow Tunnel 69
Bilsons Tree 112
Bimberi Peak 47, 49
Bird Tree 106
Birdsville 85, 124, 125, 126
Birdsville Track 43
Bishop Island 30, 31
Black Friday 126, 131
Black Head 30
Black Head and Resolution Point 95
Black Mountain 53, 54
Black Point 33
Black Rock Bridge 23
Black Springs 53, 54
Blackbutt 105
Blackheath 54
Blackwattle Bay 80
Blackwood River 77
Blanche Cup 88
Blinman 54
Blowering Dam 87
Blowfly Cave 72
Blue Cow 53
Blue Cow Mountain Terminal 58
Blue Lake 84, 86
Blue Mountains 46, 62, 65, 66, 68, 83
Blue Mud Bay 91
Bluebushes 107
Blue-green algae 103
Bluff Hill Point 31
Bluff Knoll 51, 52, 64, 65, 136
Bluff Mountain 65
bluffs 76
Boboyan Road 57
Boboyan Station 55
Bogan River 133
Bogong High Plains 47, 48, 86
Bogong West Peak 48
Bolte Bridge 80

INDEX

Bombala Line 41
Bonaparte archipelago 91
Bond Springs Station 55
Booby Island 17
Boolboonda Tunnel 68
Bool-Hack Lagoons 109
Boorara Tree 106
Booroomba Rocks 65
Booylgoo Station 127
Border Loop 68
Border Tunnel 69
Botany Bay 92
Bouganville 15
Boulia 120
Boundary Bend 36
Bourke 125
Bourke Place 24
Bowen Bridge 80
Bowen Island 98
Boxhole Crater 73
Bradshaw Station access bridge 80
Bramble Bay 79, 80
Bramble Cay 30
Bridal Veil Falls 83
Bridgewater 80
Bridgewater Bridge 81
Brindabella Ranges 83
Brindabella Road 57
Brisbane 18, 34, 118, 123, 126, 127, 128, 129, 133, 135, 137
Brisbane Bar 99
Brisbane River 79, 87, 88, 133
Broad Sound 91
Broadsound 99
Broken Hill Line 41
Bronte Park 36
Brooke 15
Brooks Bank Tunnel 58
Broome 99
Broome Cyclone 135
Broughton Island 98
Brownhill Creek Bridge 23
Bruce Highway 37
Brunswick Street Tunnel 69
Bubbler, The 88
Buccaneer archipelago 91
Buchanan Highway 43
Budidjara 13
Bulahdelah 112
Bull mallee 107
Bullarto 54, 58
Bullita Cave 72
Bulloo Lake 84
Bulloo River 78
Bunbury Bridge 81
Bunda Cliffs 66, 67
Bungendore (NSW) 31
Bungonia Gorge 63
Bunjima Drive 57
Bunya Mountains 136
Bunya Mountains National Park 68
Burakin 60
Burcher 85
Burdekin Bridge 78, 79, 81
Burdekin Delta 109
Burdekin River 77, 78, 79, 81, 87
Burdekin River delta 38
Burnie-Goodwood Siding 68
Burns 33
Bush Inn 22
bushfires 130
Bushy Park 125
Busselton Jetty 96
Bylong Tunnel 69
Byron Bay 32, 33

C

Cabramurra 53, 54, 121
Cadell Strait 98
Cadmans Cottage 19, 20
Cairns 112
Cairns-Kuranda Railway 68, 69
Caledonian Inn 22
Callide C Power Station Stack 25
Caloola Farm 37
Camballin 119
Cambewarra Mountain 50
Camooweal 33
Campbell Town Bridge 22
Canberra 34, 58, 118, 123, 124, 125, 126, 127, 128, 129, 131, 135, 137
Canberra Airport 123
Canberra Line 58
Canning Desert 43
Canning River 80
Canning Stock Route 43
canyons 76
Cape Barren Island 98
Cape Byron 30
Cape Forestier 31
Cape Howe 30
Cape Jarvis 95
Cape Lambert Jetty 96
Cape Londonderry 30
Cape Northumberland 30
Cape Pillar 95
Cape Torrens 95
Cape Tribulation 129
Cape York 30
Captain Mills Cottage 20
Captain William Hobson 17
Captains Cook Cottage 20
Carnarvon 85
Carnarvon Gorge 63, 66
Carnarvon Highway 37
Carnarvon Jetty 96
Carruthers Peak 48
Carson River Station 36, 37
Carstens 15
Cassia 103
Casterton 112
Castlereagh 18
Casuarina 103
cataracts 76
Cathedral Fig 112
Cattlewater Pass Road 57
Cauldron Pot 72
caves 71
Cazneaux Tree 112
Central Desert 43
Central Highlands 136
Central Mt. Stuart 35, 51
Central Park 24, 25
Central Plateau 47, 86
Central Tablelands 47
Cethana Dam 87
Chain of Lagoons 124
Channel Country 38, 76, 85
Channel Nine TV Tower 25
Channel Ten TV Tower 25
Channel Ten-Nine TV Tower 25
Chapel Tree 106
Charleville 118, 133
Charlotte Waters 123, 125
Charlottes Pass 53, 121, 127
Cheng Ho 14, 15
Chichester State Forest 113
Chief Street Bridge 23
Chifley Tower 24
Christmas Island 32, 33, 97, 122
Christmas Tree 106
Church of St Nicholas 21
clan 12
Clapham-Stirling West 68
Claredon Dam 88
Clarence River 80
Clarence Strait 98
Clares Bridge 22
Classification of Vegetation 102
clay plains 38
Clerk Island 30, 31
cliffs 64
Clifton Tunnel 69
climate 116
Cloncurry 125
Closed forests 104
Club Lake 86
Coalcliff 95
coast 90
coastal plains 38
Cobar 125
Cobbler Desert 43
Coburg Peninsula 17
Cockburn 33
Cockle Creek 32, 33
Cocklebiddy 125
Cocklebiddy Cave 84
Cocos Islands 30, 32, 97, 119, 125
Colindale Giant 112
Colindale Station 112
Collector 85
Collier Bay 99
Colony of South Australia 112
Commissariat Store 19, 20, 21
Commissariat Store (Norfolk Island) 20
Commo House 20
Commonwealth Avenue Bridge 80
Commonwealth Hill Station 40
Condamine River 77
Connie Sue Highway 43
Consuelo Peak 50
Consuelo Plateau 47
Cooba 107
Cook 14, 15, 17
Cook Highway 66
Cooks River 58
Cooktown 17
Coolac 85
Coolangatta 33
Cooloola 93
Coomera River Bridge 81
Coongie Lakes 109
Cooper Creek 38, 85, 112
Cooper River 76, 77, 78
Cope Road 56
Copeton Dam 87
coral coasts 97
Coral Sea Islands Territory 29
Corin Dam 88
Corin Dam Access Road

INDEX

57, 67
Corner Inlet 96
Cornish River 77
Cornwaithe Tree 106
Corra-Lynn Cave 72
Cotter River 24, 77, 88
Cotter River Valley 63
Cougal Loop 68
Coward Springs 88
Cowarie Station 37
Coxs Gap No 1 Tunnel 68
Crackenback 121, 127
Cradle Mountain 51
Crafers 22, 55
Cramsie Transmittor Tower 25
Cressbrook 20
Croom Tunnel 69
Crown Hotel 22
Cudgen road tunnel 68
Cudgewa Line 58, 68
Culgoa River 77
Cullenswood 129
Cumberland Tree 106
Cunyu Outstation 124
Currency Creek Bridge 23
Currie 33
Curtain Fig 112
Curtin Springs 121
Customs House 21
cyanobacteria 103
Cycads 104
Cyclone Ada 135
Cyclone John 135
Cyclone Rosita 135
Cyclone Steve 129
Cyclone Thelma 135
Cyclone Tracy 134, 135
Cyclone Trixie 135
Cyclone Vance 135
Cyclone Wanda 133

D

D'Entrecasteaux Channel 98
Dale Evans Bicentennial Tree 106
Dalgaranga Crater 73
Dalhousie 88
Dalhousie Springs 88
Dalton 60
Daly River 77
Daly-Reynolds Floodplain 109
Damocles Tree 106
Dampier 15, 80
Dampier-Bunbury pipeline 42
Dampier-Kambalda pipeline 42
Dandongadale Falls 83
Danger Point 31
Daniel 15
Dargo 36
Dargo High Plains Road 56
Darling Downs 39
Darling Range 136
Darling River 77, 78, 87
Dartmouth Dam 87, 88
Darwin 34, 118, 123, 126, 127, 128, 129, 135, 137
Darwin Airport 122, 123
Darwin Harbour 92
Darwin Line 58
Darwin River 87, 88
Darwin River Dam 88
Dauphin Chart 14
Davenport Ranges 29
Dawesville Bridge 80
Dawesville Cut 80
Dawson River 77
Daylesford Line 58
de Gonneville 15
de Quiros 14
de Vlamingh 15
Deakin Siding 123
Deal Island Lighthouse 95
Deer Vale 123
Delisser Sandhills 93
deltas 76
Denham 32
D'Entrecasteaux 14, 20
Derby 96, 99, 113
Derwent Bridge 85
Derwent Estuary 92
Derwent River 77, 79, 80, 81, 99
desert uplands 43
Deserts 42, 43
Devils Coachhouse 73
Devonshire Arms Hotel 22
Diamantina 97
Diamantina River 76, 78
Diamond Tree 106, 112
Dicks Tableland 47
Dieri 13
Dig Tree 112
Diggers Creek 53
Dingo Dell 53
Dingo Fence 40
Dinner Plain 53
Dirk Hartog 15, 30
Dirk Hartog Island 98
discovery and exploration 14
discovery and settlement 12
Docker River 33
Dorrigo 129
Douglas 88
Dr Hawkesworth 14
Dreaming 12, 13
Dromana 131
droughts 132
dry sclerophyll forest 105
du Fresne 15
Duckweed 106
Duncan Road 43
Dundas Strait 98
Duntroon 119, 123, 124
Duntroon Dairy 20
Duntroon House 20
Duyfken 14
Dwellingup 131

E

Eagle Rock Falls 83
Eagles Nest Cave 72
early European settlement 16
East Evelyn Road 56
East Gippsland 83
East Maitland 133
East Princes Highway 37
Eateringinna River 77
Ebor 53, 54
Eco Beach 135
Eddystone Point 119
Edels 15
Edgecombe Bay 92
Ediowie Gorge 63
Eighty-Mile Beach 93
El Grande 106
El Nino 118
El Questro Station 88
Elizabeth Downs Station 129
Elizabeth Farm 19, 20
Elizabeth River 81
Elizabeth River Bridge 81
Ellensborough Falls 83
Elliot 85
Elphinstone 69
Emu Bay Line 58
Emu Bottom 20
Encounter Bay 14
Endeavour River 17
Endeavour Strait 98
Entally House 19
Eora 13
Eraring Power Station Stack 25
Eremophila 103
Erldunda Station 124
Eromanga 35, 36
Esperance 94
Etheridge Ridge 48
Eucalyptus 103
Eucla 125
Eucumbene Dam 87
Eucumbene River 87
Eucumbene-Snowy Tunnel 87
Eucumbene-Snowy/ Snowy-Geehi/Murray One Pressure Tunnel 87
Eulo 88
Eureka Tower 24, 25
Evandale Road 41
Evelyn 56
Evelyn Central 54, 55
Exit Cave 72
Exmouth Gulf 91
Exmouth Peninsula 97
Explorers Tree 112
extreme rain events 129
extreme temperatures 124
Eyre 128
Eyre Highway 32, 37, 41

F

Falls Creek 53
Falls Creek SEC 129
Faulconbridge 55
Federation Peak 51, 65
Ferguson Tree 106
fernlands 108
Ferntree 55
Finke 124, 126
Finke River 78
firestick farming 109
Fitzroy Falls 82, 83
Fitzroy Gardens 20
Fitzroy River 78
Flinders 14, 16
Flinders Highway 37
Flinders Island 98
Flinders Ranges 38, 46, 62, 83, 105, 136
Flinders River 77, 78
Flinders Street Station Bridge 23
Flooded gum 105
floodplains 38, 76
floods 133
folding 46
forest 102
Forrest 123
Fort Denison 99
Fort Dundas 17, 20
Fortescue Bridge 80
Fortescue River 77
Forth River 87
Foundation Tree 112
Four Aces 106, 112
Fraser Island 84, 93
Frederick Meredith 18
Freestone Point Road 68

INDEX

Fremantle 99
French Island 98
French Line-QAA Line 43
Frenchmans Cap 51, 63, 65
Frobisher Map 14
Furneaux 14, 15
Fury Gorge 62

G

Gadudjara 13
Gagudju 13
Gammon Ranges 65
Gandalfs Staff 106
Garden Island 18, 80
Garie 95
Gascoyne River 76, 77, 78
Gateway Bridge 70, 79
Geehi Dam 67
Geehi River Valley 63
Geeveston Tree 106
George IV Inn 22
Georgetown 18
Georgina River 76, 78
Geraldine lead mines 24
Geraldton 99
Gerritsz 15
Ghan Line 58
Giant Stringybark 106
Giant Tingle Tree 112
Gibb River Road 43
Gibson Desert 42
Gillies Highway 66, 67
Ginini Falls 83
Gippsland 112
Gippsland Lakes 92, 93
Gladesville Bridge 70, 80
Glasshouse Mountains 59, 65
Glen Helen 35
Glen Innes 54
Glen Osmond 24
Glencoe 54
Glenelg 112
Glenelg River 77
Gloucester Tree 106, 112
Gobba Bridge 79
Golden Highway 68
Goldfields Highway 37
Gondwana 28
Gondwana Legacy 28
Goolwa River 80
Goonyella Line 68
Gordon Dam 87, 88
Gordon Falls 83
Gordon River 77, 87, 88
gorges 62, 76
Gosse Bluff 73
Goulburn 18

Goulburn River 77, 87
Government House 20, 21
Governor Head 66
Governor Hunter 21
Governor Macquarie 16
Governor Philip Tower 24
Governor Phillip 21
Governors Dairy Cottage 19
Grand Arch Cave 68
Grandis Tree 106, 112
Grant 14, 54
Granton 24
Grass tree 104
Grass tree (Kingia australis) 103
Grass tree (Xanthorrhoea australis) 103
Great Artesian Basin 87, 88
Great Barrier Reef 28, 97
Great Central Road 43
Great Cumbungi Swamp 109
Great Dividing Range 28, 29, 106, 117, 134, 136
Great Eastern Highway 37
Great Lake 85, 87
Great Northern Highway 37, 57, 58
Great Nowranie Cave 73
Great Ocean Road 66
Great Sandy Desert 42
Great Sandy Region 84, 93
Great Sandy Strait 98
Great Victoria Desert 42
Great Western Tiers 65
Green mallee 107
Greenvale Dam 88
Greenvale Line 68
Gregory 112
Gregory Tree 112
Grevillea 103
Grimes 16
Groote Eylandt 98
Grose River Valley 63
Growlling Swallet 72
Gudgenby River 77
Gudgenby Station 55, 121
Guildford 127
Guilford 58
Gulf of Carpentaria 38, 91, 96, 97
Gulf Plains 39
Gulf Track 43
gulfs 91
gullies 76
Gunbarrel-Heather

Highways 43
Guthega 53
Guyra 53, 54
Gwydir 133
Gwydir Highway 68
Gwydir River 38, 87
Gwydir River Wetlands 109

H

Haast Bluff 35
Hakea 103
Hall district 31
Hallett 56
Hamelin Pool 99, 103
Hamersley Iron Line 68
Hamersley Ranges 47, 65, 83
Hamilton Downs Access Road 57
Hanwood Bridge 80
Harris Park Viaduct 79
Harveys Return 95
Hastings 88
Hawker 60
Hawkesbury River 94
Hawthorn Railway Bridge 23
Hay Plains 39, 107
Healesville 106
Heard Island 59, 90, 97, 121, 127, 128, 136
heath 102
heatwaves 125
Hedley Tarn 84, 86
Henbury Craters 73
Herbert River Gorge 63
Herberton Road 56
Herbig Tree 112
herbland 102
herblands 108, 111
Hermannsburg 35
Hero of Waterloo Hotel 22
Hervey Bay 60, 91
Heysen Trail 37
Heysen Tunnels 68
Highway 1 37
Hillgrove Gorge 63
hillslopes 64
Hinchinbrook Island 98
Hindmarsh Island Bridge 80
Hobart 18, 34, 118, 123, 124, 126, 127, 128, 129, 131, 135, 137
Hobart Town 18
Hogan Group 31
Hollandia Nova 14
Holy Trinity Church 20, 21
Home Island 32, 33

Hoop pine 105
Hope and Anchor Tavern 22
Horizontal Falls 99
Hornibrook Viaduct 79
Houn River 87
Houtman 15
Houtman Abrolhos 97
Houtman Abrolhos Islands 17
Howitt High Plains Road 56
Hume and Hovell Track 37
Hume Highway 37
Hume Reservoir 80
Hunter River 79, 133
Hunter Valley 24, 133
Huon pine 103, 104
huts on Legges Tor 52
Hyams Beach 92

I

Ilkulka 36
Ilkulka Roadhouse 36
Illawarra coast 96
Illintjitja-Angatja Road 57
Indian Head 95
Indian Ocean 97
Ingle Hall 19, 20
Inkerman Station 37
Innamincka 36
Innot 88
Interlaken 85
Investigator Strait 98
Ipswich 24
Isdell River 105
islands 97
Ivanhoe 124, 126

J

Jabiru 119
Jack Hills 29
Jacobsz 15
Janssonius map 14
Jansz 14, 15
Jardine River Wetlands 109
Jarrah 104
Jarrahdale 129
Java La Grande 14
Java Major 14
Java Trench 97
Jenolan Caves 71, 72
Jenolan Caves Road 66, 68
Jenolan Caves-Oberon Road 68
Jeparit 85
Jerrabomberra Creek

INDEX

Bridge 23, 81
Jervis Bay 85, 122
Jervis Bay Territory coast 96
Jervis Bay Village 33
Jim Jim Creek 65
Jim Jim Falls 83
Jollimont 20
Jordon River Culvert 22
Joseph Bonaparte Gulf 91
Juna Downs Station 55

K
Kakadu 83
Kakadu Wetlands 109
Kalamunda 55, 68
Kalgoorlie 84
Kalkadoons 13
Kalpowar Queen 106
Kalumburu 33
Kanangra Deep 62
Kangaroo Creek Dam 88
Kangaroo Island 17, 98
Karijini Drive 57
Karratha 80
Karri 104, 105, 106
Karri-with-a-Hole 112
Kata Tjuta 51
Katherine Gorge 63
Katherine River 133
Katingal 55
Katoomba 112
Katoomba Falls 83
Kauri pine 104
Keewong Station 37
Keilor Plains 59
Kellar Cellar 72
Kempton Bridge 22
Kennedy Highway 56, 58
Kerry Lodge Bridge 22
Khancoban-Kiandra Road 56
Khazad-Dum 72
Kiandra 53, 121
Kimberley 83
Kimberley coast 91, 94, 96
Kinchant Dam 88
King George Falls 83
King Island 31, 98
King Jarrah Tree 113
King River 77, 113
King Sound 91
Kingoonya 85
Kings Canyon 63
Kings Cross Loop Road 56
Kings Tableland 65
Kingston 18, 32, 33
Kinrara Crater 73
Kintore 35, 36

Kintore Road 57
Kiwirrkura 36
Knapsack Gully Viaduct 70
Koedal Boepur 32, 33
Kosciuszko Plateau 48
Kosciuszko Summit Road 56
Kowanyama 36
Kulgera 33, 121
Kulumburu 36
Kunawarritji 36
Kununurra Diversion Dam 88
Kuranda Range No. 15 Tunnel 69
Kurnai 13

L
La Austrialia del Espiritu Santo 14
La Nina 118, 133
La Pérouse 15, 17
La Trobe River 133
Lacepede Beach 93
Lachlan River 76, 77, 85
lacustrine plains 38
Lady Northcotes Canyon 63
lagoons 76
Lake Acraman 73
Lake Albina 86
Lake Amadeus 85
Lake Argyle 87
Lake Augusta Road 57
Lake Barlee 84, 85
Lake Barrine 86, 113
Lake Bullenmerri 86
Lake Burley Griffin 80, 85, 87
Lake Claire 86
Lake Condah 17
Lake Cootapatamba 86
Lake Corangamite 84, 85
Lake Cowal 85
Lake Dalrymple 87
Lake Dieri 85
Lake Eacham 86
Lake Eildon 87
Lake Eucumbene 87
Lake Eyre 29, 38, 39, 78, 84, 85
Lake Frome 84, 85
Lake Gairdner 84, 85
Lake Garnpung 85
Lake George 46, 60, 84, 85
Lake George-Dalton 60
Lake Gordon 87
Lake Highway 57, 58
Lake Hindmarsh 85
Lake Hume 87

Lake Johnson Nature Reserve 104
Lake Karli Tarn 84, 86
Lake Kerrylyn 86
Lake King 93
Lake Machattie 85
Lake Mackay 60, 84, 85
Lake Macleod 84, 85
Lake Macquarie 93
Lake Maraboon 87
Lake Margaret 122, 129
Lake Menindee 87
Lake Moore 84, 85
Lake Mountain 53
Lake Mungo 38
Lake Pedder 87
Lake Pindari 87
Lake Raeside 84
Lake Sorrel 85
Lake St. Clair 85, 86
Lake Sylvester 85
Lake Tinaroo 112
Lake Torrens 84, 85
Lake Tyrell 85
Lake Wilson 86
Lake Windemere 85, 93
Lake Wivenhoe 87
Lake Woods 85
Lake Yamma Yamma 85
lakes 84
Lakes Alexandrina-Albert-Coorong 93
Lambert Centre 35
Lamington Bridge 23
Landsborough Highway 37, 41
Lansdowne Bridge 22
Larapinta Drive 57
Larapinta Trail 37
largest Baobab in Captivity 113
Larks Quarry 28
Launceston 18, 120
Launceston Hotel 22
Lawrence Hargrave Drive 66
Lawson 112
Lead Smelter Stack 25
Learmonth 135
Legges Tor 49
Len Beadel 112
Lennox Bridge 22
Level Post Bay 85
Lewisham Viaduct 23
Limestone Plains 17
Lindsay Point 30
Little River Gorge 62, 65
Little Sandy Desert 42, 86
Liverpool 18
Liverworts 103
Llangothlin 53, 54

Loddon River 77
Longford Bridge 23
Longreach 80
Lonsdale Cottage 20
Lookout Road 57
Lord Howe Island 32, 33, 59, 97
Lord Nelson Hotel 22
Lovett Tower 25
Low shrublands 111
Lowbidgee Floodplain 109
Lower Daintree River Wetland 109
Loy Yang Power Station stack 25
Lucinda Point Jetty 96
Lyell Highway 66
Lyonville 54

M
Macarthur Bridge 79
Macassan voyagers 15
Macassans 17
Macdonnell Ranges 113
MacDonnell Ranges 46
Macgregor 36
Macintyre River 77
Mackay 128, 136
Mackay Cyclone 135
Macquarie 133
Macquarie Arms Hotel 22
Macquarie Culvert 22, 23
Macquarie Harbour 91
Macquarie Island 31, 97, 121, 123, 136
Macquarie Marshes 109
Macquarie Towns, the 18
Macumba River 77
Mahogany Creek 22
Mahogany Inn 22
Maidens gum 105
Main North Line 58
Main Range 65, 67
Main West Line 41
Maireana 103
Malangangerr 17
Malays 15
Maldon Bridge 70
Mallacoota 33
Mallee Wheatlands 16
Mandildjara 14
mangrove formations 102
Mangroves 111
Manjimup 113
Manjimup district 112
Mann Ranges 86
Marble Bar 119, 120
Marchinbar Island 95

155

INDEX

Marchinbar Island south-east coast 96
Marco Polo 14
Mardie Station 124, 135
Marianne North Tree 113
Maribyrnong River Bridge 70
Marlborough Highway 57
Marrakai Apartments 25
Marrangaroo Tunnel 69
Marrawah 33
Marree 85, 125
Mary River Floodplain 109
Maryville-Cumberland Junction Road 68
Massey Gorge 63
Mataranka 88
Mawson Peak 29, 47, 49
Mawson Trail 37
Maydena Tree 106
McBride Volcanic Province 38, 59
McCartney Street Bridge 23
McDonald Island 59, 97
Meckering 60
Meckering-Cadoux districts 60
Meeberrie 60
Meekatharra Desert 43
Melaleuca 103
Melanesians 15
Melbourne 34, 118, 123, 126, 127, 128, 129, 135, 137
Melbourne Central 24
Melrose 112
Melville 135
Melville Island 17, 20, 98
Menamatta Tarns 86
Menangle Bridge 23
Mendonca 14, 15
Menindee 85
Menindee Lake 84, 85
Mereenie Loop Road 57
Merritts Creek Bridge 70
Messmate stringybark 105
Meteorological Office 123
Mettler 121
Miena 54, 85
Mildura 119, 120, 124, 125, 126
Mill Creek Bridge 23
Millaa Millaa 123
Millers Bluff 51
Millicent 60
Millstream 88
Minjilang 33

Minnamurra River 80
Mission River 80
Missionary Plain 86
Mitakoodi 13
Mitchell grass 108
Mitchell Highway 37, 41
Mitchell River 93
Mitta Mitta River 81, 87, 88
Mitta Mitta River Bridge 81
Mokoan Dam 88
Molonglo Gorge 62
Molonglo River 87
Molonglo River Bridge 81
Monaro Highway 58
Monaro-Snowy Mountains 47
Monash Freeway Viaduct 79
Mongarlowe 102
Mongarlowe mallee 102
Monkerai Tunnel 69
Monkira Monster 113
Monkira Station 124
Montezuma Falls 83
Moolgum Bridge 70
Moomba-Sydney pipeline 42
Mooney Mooney Creek Bridge 70
Moreton Bay 17, 91, 93
Moreton Bay (Brisbane) 92
Morialta Falls 83
Mornington Island 98
Mornington Peninsula 17
Morton Bay Wetlands 109
Mosman 135
Mosman Tornado 135
Mosses 103
Motpena Station 129
Moulting Lagoon 93
Mount Buller 127
Mount Lofty 127
Mount Tree 106
Mountain ash 105
Mountain grey gum 105
Mountain Hut, the 22
Mr Jessop Tree 106
Mt Ainslie 51
Mt Alice Rawson 48, 86
Mt Anne 51, 136
Mt Augustus 51, 52
Mt Barrington 50
Mt Barrow Access Road 57
Mt Bartle Frere 47, 64
Mt Baw Baw 50, 53

Mt Beerwah 50
Mt Bennett 68
Mt Bogong 47, 48
Mt Bogong 64
Mt Bowen 49, 95
Mt Boyle 47
Mt Brown 64
Mt Bruce 49
Mt Bruce Access Road 57
Mt Bruny 49
Mt Bryan 50, 55
Mt Bryan East Road 56
Mt Buangor 50
Mt Buffalo 53, 63, 65
Mt Buffalo Plateau 47
Mt Buller 50, 53
Mt Buller Summit Road 56
Mt Burr 59
Mt Canoblas 50, 59
Mt Carbine 49
Mt Charles 49
Mt Clear 31
Mt Cobberas 48
Mt Conner 51
Mt Cope 48
Mt Cordeaux 50
Mt Dalrymple 50
Mt Dandenong 50
Mt Dare 36
Mt Donna Buang 53
Mt Dromedary Forest Preserve 108
Mt Eccles 59
Mt Edward 49
Mt Etheridge 48
Mt Exmouth 50
Mt Fainter South Peak 48
Mt Feathertop 48
Mt Field 53
Mt Frankland 51
Mt Franklin 31, 51
Mt Franklin Road 57
Mt Frederick 49, 62
Mt Gambier 59, 84, 106
Mt Gambier Airport 123
Mt Garnet 88
Mt Gee 71
Mt Gibraltar 50
Mt Giles 49
Mt Gingera 49, 64
Mt Ginini Summit Road 57
Mt Ginini Tower 52
Mt Gower 49, 95
Mt Gudgenby 51
Mt Hamilton 49
Mt Hay 51
Mt Henry Bridge 80
Mt Hotham 48, 53, 54, 121, 127

Mt Hypipamee 73, 86
Mt Imlay 50
Mt Kaputar 50, 52, 59
Mt Kelly 51
Mt Kiangarow 50
Mt King-George 50
Mt Kosciuszko 29, 47, 48, 84, 86, 105, 122, 136
Mt Latrobe 50
Mt Lee 48
Mt Lidgbird 49
Mt Lindesay 50
Mt Loch 48
Mt Lofty 50, 54, 67, 83, 136
Mt Lofty Ranges 83, 86, 105
Mt Lofty Summit Road 56
Mt Lofty-Flinders Ranges 60
Mt Macedon 50, 54
Mt Maria 49
Mt McKay 48
Mt Meharry 47, 49
Mt Morris 50
Mt Munro 49
Mt Murray 49
Mt Nameless 66
Mt Nameless 4WD track 57
Mt Nameless Tower 52
Mt Napier 59
Mt Narryer 29
Mt Nelse North Peak 48
Mt Niggerhead 48
Mt Olga 51
Mt Ord 51
Mt Ossa 29, 47, 49, 86
Mt Pelion West 49
Mt Razorback 51
Mt Remarkable 50
Mt Robinson Rest Area Access Road 57
Mt Schank 59
Mt Selwyn 53
Mt Sonder 51
Mt St Gwinear 53
Mt Stevenson 51
Mt Stirling 53
Mt Surprise 88
Mt Tambourine 104
Mt Tempest 93
Mt Townsend 48, 64, 86, 136
Mt Twynam 48
Mt Wallum 57
Mt Warning 50, 59
Mt Waverley 20
Mt Wellington 50, 51, 120, 121, 123, 127
Mt Whinham 50
Mt William 50

INDEX

Mt Woodroffe 47, 49
Mt Wycheproof 52
Mt Zeil 47, 49, 64
Mulga 107
Mulka Bore 123
Mundi Mundi Plain 39
Mungeranie Bore 123
Murchison River 77, 78
Murchison River Gorge 63
Murnepowie Station 124
Murphys Creek-Harlaxton 68
Murray 16
Murray Mallee 28
Murray River 31, 77, 78, 80, 87
Murray River Bridge 81
Murrumbidgee River 76, 77, 79, 80, 81
Musgrave Ranges 49
Mustoms Creek Bridge 23
Myall Creek Bridge 23
Mylyie Station 123

N

Naas River 77
Namoi 133
Narrenjeri 13
Narrogin 127
Neales River 77, 81
Neds Corner 123, 124
Neerim 112
Nepean River 79
Nepean River Bridge 70
Neurragully Waterhole 113
New England 47
New England Highway 37
New Norfolk 18
New South Wales 91, 97, 125
New Southern Railway Line 58
New Year Island 31
Newcastle 18, 60
Newcastle Bight Beach 93
Newell Highway 37
Newnes Line 58, 68
Newstead House 20
Ngarrindjeri 13
Ngdadjara 13
Nhulunbuy 32, 36
Niggly Cave 72
Nimmitabel 54
Ninety-Mile Beach 93
Ninety-Mile Desert 107
Ningaloo Reef 97
Noble Tree 106

Nogoa River 87
Noosa Lakes 93
Noosa River Wetlands 109
Norfolk Island 18, 30, 32, 97
Norfolk Island pine 105
Normanton 119
North Coast Line 68
North Esk River 67
North of Cairns coast 96
North Pole Well 29
North Rams Head 48
North Shore 135
Northbourne Avenue 41
Northbridge 'Suspension' Bridge 70
North-East Highlands 59
north-eastern Tasmania 133
north-eastern Victoria 133
Northern Tablelands 47, 83
Northern Territory 91, 97
North-West Coastal Highway 37, 41
North-West Plains 39
Nowra 113
Nullarbor 95
Nullarbor coast 96
Nullarbor Plain 28, 38, 39, 71, 84, 107
Nuyts 15
Nuytsland 14
Nyang Station 120
Nyngan 118

O

Oberon 54
Ocean Beach 93
Oenpelli 121
Old Adaminaby 53, 54
Old Government House (Parramatta, NSW) 19, 21
Old Government House (Western Port, Vic) 20
Old Government Stores 21
Old Homestead Cave 72
old log beam bridge 23
Old Spot Hotel 22
old stone culverts 22
Olga Gorge 62, 65
Olinda 55
Olsens Lookout 67
Omega Tower 25
Omeo 54
Omeo Highway 58
Oodnadatta 88, 119, 120, 123, 124, 125, 126

Oodnadatta Meteorological Office 123
Oodnadatta Track 41, 43, 58
Open forests 105, 111
Open heaths 111
Open herblands 111
Open scrubs 111
Ophthalmia Range 83
Orbost 81
Orchidaceae 103
Ord Dam 88
Ord River 87, 88
Orroroo 113
Otford Tunnel 69
Otway coast 96
Outback Highway 37
Ouyen 124
Overland Track 37
Owen Springs 20
Oyster Bay 91

P

Pacific Highway 37
Pacific Ocean 97
Palmerston 18
Papunya 35
Parafield 127
Paralana 59, 88
Paralana Fault 60
Paralana Hot Springs 66
Pardoo 58
Pardoo Roadhouse 135
Parramatta 18
Parramatta River 80
Patjarr 35, 36
Patrick Taylor Cottage 20
Paynes Find 85
Peak Charles 51
Peak Range 59
Pebbly Beach 92
pediplain 38
Pedirka 42
Pedra Blanca 30
Peel-Harvey Inlet 93
Pekina Creek River Red Gum 113
Pelsart 15
Pemberton 106, 112, 113
Pencil pines 104
peneplain 38
Pennant Hills 127
Perisher Gap 53
Perisher Valley 53
Perth 34, 118, 123, 126, 127, 128, 129, 135, 137
Perth Bridge 22
Peterborough 66
Pheasant Nest Bridge 70
Picton 60
Picton-Mittagong Loop

Line 68
Pidgeon-House 50
Pieman River 87
Pieterszoon 15
Pilbara 88, 135
Pine Creek Hotel 22
Pine Lake Dam 88
Pine Range No. 2 Tunnel 69
Pinjarra-Dwellingup 68
Pinnacles Desert 43
Pinnacles Road 57
Pintupi 13
Pioneer River 80
Pipalyatjara 33
Pitjantjatjara 13
Pitt Town 18
Pitt Water 80
Pitt Water Estuary 80
plain 38
Plenty-Donahue Highways 43
plunge pool 82
Poaceae 103
Poatina Highway 57
Poeppels Corner 60
Point D'Entrecasteaux 97
Point Danger 30
Point Lookout 50, 65
Point Lookout Road 56
Point Peron 80
Porcupine grass 108
Port Adelaide 99
Port Albert Hotel 22
Port Campbell 94
Port Campbell Coast 96
Port Darwin 99
Port Darwin Wetlands 109
Port Essington 17
Port Germain Jetty 96
Port Hedland 80, 92
Port Macdonnell 33
Port Phillip Bay 91
Port Phillip Bay (Melbourne) 92
Port Phillip Heads 99
Port Pirie-Peterborough Line 58
Portland 18
Portuguese explorers 15
Precipitous Bluff 51, 64
Pretty Valley Pondage 86
Pretty Valley Pondage Road 56
Prince Alfred Bridge 80
Princes Highway 37
Princes Highway West 37
Princetown 66
Prison Tree 104, 113

INDEX

Pryde Creek 88

Q
Queensland 91, 97
Quobba Point 32

R
Rabbit Flat Roadhouse 36
Rabbit Proof Fence No.1 40
Rabbit Proof Fence No.2 40
Raffles Bay 17
Railway Station 21
Rainforests 111
Rams Head 48
Rams Head North 48
rapids 76
Rattlesnake Point 95
Ravenshoe 54
Ravenshoe Line 58
Rawnsley Bluff 50
Recherche Bay 20
Red cedar 105
Red tingle 105
Redbank Mine 129
Redcliffe camp 17
Redcliffe Viaduct 79, 80
Reece Dam 87
Reeves Point 17
Rendezvous Creek 65
Restdown Bridge 22
Rhydaston Tunnel 69
Rialto Towers 24
Richmond 135
Richmond (NSW) 18
Richmond Bridge 22, 23
Rig Road-K1 Line-Warburton Track 43
Riparian Plaza 25
Risdon Cove 17
River cooba 107
River Light 77
River red gum 104, 106
Riverina region 38
riverine 38
Robertson-Bowral 60
rock coasts 94
rock plains 38
rockholes 76
Rockingham 80
Rodondo Island 94
Roebuck Bay 91
Roma 136
Roper Bar 120
Roper River 77, 78
Roper Valley 129
Roseberth Station 123
Rosebery 83
Rosella-Yandicoogina Line 58
Ross 123
Ross Bridge 22
Ross Camm Bridge 80
Ross River 35, 88
Ross River Dam 88
Rough tree fern 108
Round Mountain 50
Round-leaved gum 105
Royal Bulls Head Inn 22
Royalla railway station 58
Rubicon timber mill 131
Russel Tarn 86
Russell Falls 83

S
sacred sites 12
Sale-Sydney pipeline 42
Salt marshes 111
Saltbushes 107
Saltlake 84
Sandgate 127
Sandover Highway 43
sandplains 38
Sandy Creek North Branch 88
Sandy Point 32
Santa Teresa 35
Santos House 25
Sawpit Creek 53
Scenic Railway 69
Schlink Pass 67
Scottish Presbyterian Church Ebenezer 21
scrub 102
scrubs and heaths 107
Sea Lake 85
Seagrasses meadows 111
Seamans Hut 52
Sedgelands 108
Seisa 32, 33
Seldom Seen store 54
Serendipity 72
Serpentine Falls 83
Serpintine River 87
Serviceton 33
settlement 12
Seventy-Five Mile Beach 93
Severn River 87
Seymour Street 68
Shannon 127
Shannon River 87
Shannonvale Station 37
Shark Bay 91
Sheehan Bridge 79
Shell beach 92
Shelley 58
Sheoak Road 68
shield plains 38
Shining gum 105
Shoalhaven Gorge 63
Shoalwater Bay 91
Shooters Hill 53
shrubland 102
shrublands 107, 111
significant Australian droughts 132
Silver Lake 86
Simpson Desert 38, 42
Sirius 18
Skitube rack railway 68
Skull Tanks 84
Sleeps Hill Tunnel 69
Smiggin Holes 53
Smiggin Holes-Guthega Link Road 56
Smokers Gap 53
Snow gums 105
Snowy Mountains 48, 86, 87, 88, 105, 136
Snowy Mountains Highway 58
Snowy River 105
Snowy River Trestle Bridges 81
Snowy River Valley 63
Sorell Causeway Bridge 80
South Australia 91, 97, 125
South East Cape 30, 31
South Esk River 77
South Geelong Tunnel 69
South Head 129
South Johnstone Experimental Station 123
South Johnstone Post Office 123
south of Perth 99
South Para 87
South Para River 87
South West Island 30
Southern Highlands 83
Southern Ocean 97
Southport 32
Sparkes Gully 73
Spencers Creek 136
Spencers Gulf 14, 62, 91
Spinifex 108
Spion Kopje 48
spits and sand dunes 93
Split Yard Creek 88
Splitters Creek Bridge 23
spot fire 130
Spotted gum 105
Sprigg River 105
Spring Creek Bridge 70
Spring Tidal Ranges 99
Springfield 20
Springleigh Bore 87
Springton 112
St Allouarn 15
St James Old Cathedral 21
St Johns 21
St Marys Peak 50, 65
St Mathews 21
St Stephens 21
St Vincent Gulf 14, 62, 91
St Vincents Gulf (Adelaide) 92
Standley Chasm 63
Stanley 33, 54
Stanthorpe 127
Stanwell Creek Viaduct 70
Stanwell Stack 25
Steamers Beach 66
Steamers Head 95
Steavensons Falls 83
Steep Point 30
Stewart Island 30
Stewarts Karri 106
Stirling 121, 122
Stirling Ranges 46
Stirling Ranges Drive 66
Stockton Bridge 79
Stokes Hill Lookout Road 56
Stonor 58
stony plains 38
storms 134
Story Bridge 70, 79
Strahan 126
straits 98
Strathgordon 36
Strawberry Hill Farm 20
streams 76
Stromatolites 103
Strzelecki Creek 85, 125
Strzelecki Desert 42
Strzelecki Peak 49
Strzelecki Track 43
Stuart Highway 37, 58
Sturt 125
Sturt Highway 37, 41
Sturt Tree 113
Sturts Stony Desert 38, 42
Subiaco Rail Tunnel 69
Sullivans Bay convict camp 17
Summit Hut 52
Summit Road 56
Sunday Strait 98
Sunnywest Farm 122, 123
Surveyor-General Hotel 22
Sutton 60

INDEX

swamp mahoganies 104
Swampy Plain Bridge 67
Swan River 80, 81
Swan River Bridge 80
Swan View Tunnel 69
Swan-Canning Estuary 92
Swanport Bridge 80
Swansea 128
Sydney 18, 34, 118, 123, 126, 127, 128, 129, 131, 135, 137
Sydney blue gum 105
Sydney Hailstorm 135
Sydney Harbour 58, 81, 94
Sydney Harbour 92
Sydney Harbour Bridge 70, 79, 81
Sydney Town 18

T
tablelands 46
Tacky Creek Bridge 22
Talbingo Dam 87, 88
tall open forests 105
Tallaroo 88
Tallowwood 105
Tallowwood Point 129
Tamar Estuary 77
Tamar River 80
Tanami Desert 38, 42
Tanami Downs Station 37
Tanami Track 43
Tanbryn 129
Tasman 15
Tasman Arch 94
Tasman Bridge 70, 79, 80
Tasman Peninsula 66
Tasman Peninsula coast 96
Tasmania 91, 97, 98, 105, 107, 108
Tasmanian blue gum 105
Tasmanian Highlands 136
Tatra Inn 53
Tennant Creek 60
Tennant Creek district 60
Terminalia 103
Terra Australis 14
Terra del Zur 14
terraces 76
Terragong Swamp Bridge 80
Tharwa 18, 36
The Big Hole 73
The Big Tree 106
The Bottomless Pit 72
The Bush 102
The Coastline 90
The Coorong 92
The Crater 73, 86
The Desert 43
The Gorge 63, 65
The Gorge, Mt Lofty Ranges 65
The Grange 20
The Grotto 73
The Grotto 86
The Horn 50
The Kimberley 105
The Little Desert 43
The Oaks 20, 21
The Pilbara 38
The Queenslander 72
The Residency 20
The Rip 99
The Round House 20, 21
The Sky 116
The Springs 127
The Twelve Apostles 95
The Windmill 20
Theda Station 36
Thermal Pool 88
Thijssen 15
Thomson Dam 87
Thomson River 77, 80, 87
Thornton Peak 50
Thorpdale 106
Thredbo 53, 54
Thredbo-Crackenback 53, 54
Thursday Island 32
Tibooburra 36, 119, 120
tidal plain coasts 96
tidal plains 38
Tidal River 32, 33
Tidbinbella Peak 51
tides 99
Tilmouth Crossing 35, 36
Timber Creek 112
Tirari Desert 42
Ti-Tree hotel 35
Tiwi 13
Tjuntjuntjara 36
Todd River 133
Tolmie 54
Tom Price 54
Tom Price-Paraburdoo Road 57
tombolo 93
Tomewin 122
Toowoomba 59
Top End 88
Torbay Head 30
Torrens 54, 55
Torrens River 77, 88
Torres 15
Torres Strait 98
tower above Heavitree Gap 52
Tower Hill 59, 84
Towrang Creek Viaduct 22
Trans-Continental Line 41
Travellers Village 33
Treasure's Homestead 55
Tree ferns 104
Tree of Knowledge 113
Trentham 54
Tropic of Cancer 117
Tuena Hotel 22
Tully 122
Tully Falls 83
Tumblong deviation 68
Tumoulin 58
Tumoulin Road 56
Tumut Pond Dam 87
Tumut River 87, 88
Turramurra 129
Tussock grasslands 108
Tweed Heads 33
Twin Ghost Gums 113
Twin Kauris 113
Twin Peaks 51
two stone bridges 22

U
Uluru 51, 136
Undara Lava Tubes 71, 72
United Services Hotel 22
unnamed fig 113
unnamed small-leaved fig 113
Unvegetated areas 111
Uriarra Camp 120
Uriarra Forestry Camp 33
Useless Loop 32, 33

V
valleys 62
van Colster 15
Van Diemen Gulf 91
Van Diemens Land 14, 18
van Hillegom 15
Veevers Crater 73
vegetation 112
Victoria 91, 97, 125
Victoria River 77, 78, 80, 99
Victoria River Downs 40
Victorian Alps 48, 86, 136
volcanic plains 38
volcanoes 59

W
Waddywood 103
Wagga Wagga Viaducts 81
Walcott Inlet 99
Wallaman Falls 65, 83
Wallangarra 33, 127
Wallaroo 58
Walls Cave 71
Walls of China 38
Walpiri 13
Walpole 105, 112
Walshs Pyramid 52
Wanaaring 36
Wandering 121
Wangkatja 13, 14
Wanman 13
Wapet Road-Gary Junction Road 43
Warakurna 112
Waratah 121, 123
Warburton 36, 112
Warburton River 78, 85
Warburton Road 43
Warragamba Dam 87
Warrego Highway 37
Warrego River 133
Warrnambool 14
Warrumbungles 59
Warwick 127
Waterfalls 76, 82
waterholes 76
Watsons Crags 65
Watts River 106
Wayatinah Tree 106
weather 116
weather changes 116, 117
weathering 46
Weeaproinah 123
Weipa 80
Wentworth 113
Wentworth Falls 83
Werribee River Viaduct 70
Wesleyan Methodist Church 21
West Bardon 55
West Gate Bridge 70, 79, 80
West Gate Freeway Viaduct 79
West Wallabi Island 20
Western Australia 91, 97, 99
Western Desert 43
Western Distributor Viaduct 79
Western Districts 38, 59
Western New South Wales 131
Western Port 92

wet desert *43*
wet sclerophyll forests *105*
wetlands *108*
Whim Creek *129, 135*
Whipstick mallee *107*
White Cliffs *125*
White mallee *107*
Whitfield *83*
Whitsunday Passage *98*
Wihareja Homestead *55*
Wilberforce *18*
Wilcannia *124*
Wilga Mia *24*
Willandra Creek *76*
Williamsdale *33, 54*
Willis Island *120, 124*
Willochra Plain *39*
willy-willies *134*
Wilpena Pound *112*
Wilsons Peak *52*
Wilsons Promontory *30, 94*
Wilsons Valley *53*
Windjana Gorge *63*
Windmill Cutting *68*
Windsor *18*
Wingelinna *33*
Wingen *59*
Winnellie *55*
Wintawata-Ulkiya Road *57*
Winton Swamp *88*
Wire Plain *53*
Wiso Road *43*
Wittenoom *126*
Wittenoom Road *67*
Witts *15*
Wivenhoe Dam *88*
Woakwine Cutting *68*
Wolfe Creek Crater *73*
Wolfe Rock *30*
Wolindilly River *87*
Wollemi National Park *113*
Wollemi Pines *113*
Wollomombi Falls *65, 83*
Wollomombi Gorge *63*
Wollongong *129, 133*
Wombeyan Caves Road *66, 68*
Wonboyn *33*
woodland *102*
woodlands *106, 111*
Woolmers *20*
Woolnorth Point *31*
Woolpack Hotel *22*
Wooramel Seagrass Bank *109*
Woronora Bridge *70*
Woy Woy Tunnel *69*
Wreck Bay *31*
Wrest Point Hotel *25*
Wulgulmerang Creek Falls *83*
Wurdee Buloc Dam *88*
Wycheproof *52*
Wyelangta *122*
Wyndham *86, 99, 119, 104, 113*
Wytfliet Map *14*

Y

Yam Creek mines *24*
Yambarran Range *66*
Yandama Downs *124*
Yandama Downs Station *123*
Yarra River *79, 80*
Yarran *107*
Yarrongobilly *88*
Yelgun-Chinderah Freeway *68*
Yelta *33*
Yongala *127*
York Sound *91*
Youanmi *85*
Younghusband Peninsula *93*
Yulara *85, 127*
Yungaburra *112*
Yuroke Creek *88*
Yurr Yurr *36*

Z

zamia palm *104*
Zebedee *88*
Zero Tower *25*
Zig Zag *68*
Zig Zag No 10 Tunnel *69*
Zuytdorp Cliffs *95*